Lecture Notes in Mathematics

Edited by A. Dold and B. Eckmann

949

Harmonic Maps

Proceedings of the N.S.F. – C.B.M.S. Regional Conference,
Held at Tulane University, New Orleans
December 15 – 19, 1980

Edited by R. J. Knill, M. Kalka, and H.C.J. Sealey

Springer-Verlag
Berlin Heidelberg New York 1982

Editors
Ronald J. Knill
Morris Kalka
Department of Mathematics, Tulane University
New Orleans, LA 70118, USA

Howard C. J. Sealey
Department of Mathematics, University of Utah
Salt Lake City, UT 84112, USA

AMS Subject Classifications (1980): 53-06, 53 C 05

ISBN 3-540-11595-1 Springer-Verlag Berlin Heidelberg New York
ISBN 0-387-11595-1 Springer-Verlag New York Heidelberg Berlin

Printing and binding: Beltz Offsetdruck, Hemsbach/Bergstr.
2146/3140-543210

PREFACE

These proceedings report the substance of talks and papers contributed by participants in the N.S.F.-C.B.M.S. Regional Conference on Harmonic Maps at Tulane University Dec. 15-19, 1980. The principal lecturer at that conference was James Eells of the University of Warwick. His lectures, as is customary with such conferences, will be published separately in the blue series of CBMS Regional conference reports. That report was co-authored by Luc Lemaire.

The format of the conference was ten lectures by Eells, ten lectures by selected participants. These latter lectures along with two contributed papers occur here. Thus the Eells-Lemaire report and these lectures are to be regarded as companion volumes. The Eells-Lemaire work carefully lays down the foundation for the formalism, in the context of differential geometry, necessary for the development of the theory of harmonic maps, and systematically applies that formalism to selected topics.

These proceedings concern related results in the area of harmonic maps. The two volumes together are not exhaustive of the current state of the theory, however they represent the recent efforts of some of the leading contributors to its development.

The editorial committee would like to acknowledge first of all James Eells for the selfless hard work and preparation that went into his lectures, for his leadership, and his overall good nature which contributed to the pleasant and stimulating exchanges at the conference.

We would also like to use this opportunity to thank Ms. Jackie Boling whose administrative expertise kept the conference running smoothly, Mrs. Hester Paternostro who ably handled all correspondence for the conference, and Mrs. Phuong Q. Lam for her efficient help in editing and retyping several of the papers occuring here.

The editorial committee was chaired by Ronald J Knill who would like to personally acknowledge the contribution of M. Kalka, H. C. J. Sealey, A. L. Vitter and P.-W. Wong for their expert advice and support throughout.

We would all like to acknowledge the financial support of the National Science Foundation for the conference. In addition the editors wish to acknowledge Tulane University for substantial financial and staff support without which these proceedings would not have been produced.

TABLE OF CONTENTS

MILNOR NUMBER AND CLASSIFICATION OF ISOLATED SINGULARITIES OF HOLOMORPHIC MAPS

by

Bruce Bennett & Stephen S.-T. Yau

§ 0. Introduction

A map $f: M \to N$ between Riemannian manifolds is said to be harmonic if it is a critical map for the energy functional $E(f)$ (cf. § 1). In [4] Eells-Sampson proved the following fundamental theorem. If M and N are both compact and N has nonpositive sectional curvature, then every continuous map from M to N is homotopic to a harmonic map. Hartman [7] proved that the harmonic map is unique in each homotopy class if N has strictly negative curvature.

For more detail theory of harmonic maps, the reader should consult the excellent survey article [3] by Eells and Lemaire. Perhaps one of the interesting and difficult problems is the classification of singularities of harmonic maps between complex Kähler manifolds. In [26, 27] J. C. Wood gave a complete classification of singularities of harmonic maps between surfaces (real dimension). It is well known that a holomorphic or conjugate holomorphic map between Kähler manifolds is always harmonic. The problem of determining under what conditions the converse holds is an important and difficult problem. Recently Siu [24] has defined a notion of strongly negative curvature tensor and proved the following important theorem.

<u>Theorem (Siu)</u> Suppose M and N are compact Kähler manifolds and the curvature tensor of N is strongly negative. Suppose $f: M \to N$ is a harmonic map and the rank over $\mathbb{R}$ of the differential df of f is at least 4 at some point of M. Then f is either holomorphic or conjugate holomorphic.
In view of these facts, we shall only discuss singularities of holomorphic maps in this paper. At least this is the first step towards understanding singularities of harmonic maps. Actually only isolated singularities of holomorphic maps will be considered. Under these assumptions, the techniques

from algebraic geometry and complex analysis can be brought in.

Let $f: U \subseteq \mathbb{C}^3 \to \mathbb{C}$ be a holomorphic function on an open neighborhood U of 0 in $\mathbb{C}^3$. A point x in U is called a singular point of a map f if the complex gradient ∇f is zero at x. We assume that $0 \in U$ is an isolated singularity of f and $f(0) = 0$. In [2], [13], and [30], the theory of classification of singularities was quite well developed by using the geometric genus of the singularity. In this paper, we shall classify singularities according to Milnor number. Although the later classification turns out not as natural as the former one, in practice the later one may be more useful because the Milnor number is easier to compute than the geometric genus. The later approach was due to Mather [15], Siersma [21], [22] and Arnold [1]. In § 1, we recall some of their results as well as Milnor's results on topology of hypersurface singularities. In § 2 we give a formula which relates the Milnor number and the invariants of any resolution of the singularity (cf. Theorem A and Theorem B). These results were obtained some years ago and have been informally circulated to some extent. It seems to us that there is still interest in this article. In § 3 we develop certain results related to sheaves of p-forms at isolated hypersurface singularities.

We would like to thank P. Griffiths, H. Hironaka, M. Kuga, J. Milnor, D. Mumford, H. Laufer and Y.-T. Siu for discussions related to this work.

§ 1. Isolated singularities of holomorphic maps

Suppose $f: M \to N$ is a map between two Riemannian manifolds with Riemannian metrics

$$ds_M^2 = g_{ij}dx^i dx^j$$

$$ds_N^2 = h_{\alpha\beta}dy^\alpha dy^\beta$$

(Here the summation convention is used.) We can define $\operatorname{trace}_{ds_M^2} = f^* ds_N^2$ at $p \in M$ as the sum of the critical values of

$$\frac{f^* ds_N^2}{ds_M^2}$$

on the set of all nonzero tangent vectors of M at p, with each critical value counted as many times as the dimension of its associated critical set. It is clear that

$$\operatorname{trace}_{ds_M^2} f^* ds_N^2 = g^{ij}(h_{\alpha\beta}\cdot f)\ \frac{\partial f^\alpha}{\partial x^i}\ \frac{\partial f^\beta}{\partial x^j}$$

We define the energy density $e(f)$ of f by

$$e(f) = \frac{1}{2}\operatorname{trace}_{ds_M^2} f^* ds_N^2$$

and the energy $E(f)$ of f by

$$E(f) = \int_M e(f)$$

Definition: A map f: $N \to M$ between Riemannian manifolds is said to be harmonic if it is a critical map for the energy functional $E(f)$.

The Euler-Lagrange equation for the energy functional $E(f)$ is

$$(*) \qquad g^{ij}\left[\frac{\partial^2 f^\alpha}{\partial x^i \partial x^j} - {}^M\Gamma^k_{ij}\frac{\partial f^\alpha}{\partial x^k} + {}^N\Gamma^\alpha_{\beta\alpha}\frac{\partial f^\beta}{\partial x^i}\frac{\partial f^\alpha}{\partial x^j}\right] = 0$$

where ${}^M\Gamma^k_{ij}$ and ${}^N\Gamma^\alpha_{\beta\alpha}$ are respectively the Christoffel symbols of M and N.

From now on we assume that M , N are Kähler manifold with Kähler metrics $g_{i\bar{j}}dw^i d\overline{w^j}$ and $h_{\alpha\bar{\beta}}dz^\alpha d\overline{z^\beta}$ respectively. For any smooth map f: $M \to N$ let

$$f^\alpha_i = \frac{\partial}{\partial w^i} f^\alpha$$

$$f^\alpha_{\bar{i}} = \frac{\partial}{\partial \overline{w^i}} f^\alpha$$

Then equation (*) becomes

$$g^{i\bar{j}}\left[\frac{\partial^2 f^\alpha}{\partial w^i \partial \overline{w^j}} + {}^N\Gamma^\alpha_{\beta\alpha} f^\alpha_i f^\beta_{\bar{j}}\right] = 0$$

Hence a holomorphic or conjugate holomorphic map between Kähler manifolds is always harmonic.

Let f: $M \to N$ be a holomorphic map. We shall always assume $n = \dim M > \dim N = k$. A point x in M is said to be a singular point of the map f if the Jacobian of f at x is not of maximal rank. In this paper, we shall assume that f has only isolated singularities. Clearly $V = f^{-1}(f(x))$ is a local complete intersection at x . As far as the classification of isolated singularities is concerned, the problem is local. We may assume from now on that $M = \mathbb{C}^n$ and $N = \mathbb{C}^k$. The classification theory works particularly well for the case $k = 1$.

Two germs of smooth functions $f: (\mathbb{C}^n,0) \to (\mathbb{C},0)$, $g: (\mathbb{C}^n,0) \to (\mathbb{C},0)$ are equivalent, if they belong to the same orbit of the group of germs of holomorphic diffeomorphisms $(\mathbb{C}^n,0) \to (\mathbb{C}^n,0)$. Two germs $f: (\mathbb{C}^n,0) \to (\mathbb{C},0)$ and $g: (\mathbb{C}^m,0) \to (\mathbb{C},0)$ are stably equivalent, if they become equivalent after a direct addition to both of nondegenerate quadratic forms (e.g., $f(x) = x^3$ is stably equivalent to $g(x,y) = x^3 + y^2$ but not to $h(x,y) = x^3$). Two stably equivalent germs on equidimensional spaces are equivalent. Clearly it suffices to classify singularities up to stable equivalence. Before we do that, let us first recall Milnor's results on the topology of hypersurface singularities.

Let $f: U \subseteq \mathbb{C}^{n+1} \to \mathbb{C}$ be an analytic function on an open neighborhood U of 0 in $\mathbb{C}^{n+1}$. We denote

$$B_\varepsilon = \{z \in \mathbb{C}^{n+1}: \|z\| \leq \varepsilon\}$$

$$S_\varepsilon = \partial B_\varepsilon = \{z \in \mathbb{C}^{n+1}: \|z\| = \varepsilon\}$$

Then:

<u>Theorem 1.1</u> For $\varepsilon > 0$ small enough the mapping $\varphi_\varepsilon: S_\varepsilon - \{f = 0\} \to S^1$ defined by $\varphi_\varepsilon(z) = f(z)/|f(z)|$ is a smooth fibration.

<u>Theorem 1.2</u> For $\varepsilon > 0$ small enough and $\varepsilon >> \eta > 0$ the mapping $\psi_{\varepsilon,\eta}$: $(\mathrm{Int}\ B_\varepsilon) \cap f^{-1}(\partial D_\eta) \to S^1$ defined by $\psi_{\varepsilon,\eta}(z) = f(z)/|f(z)|$, where $\partial D_\eta = \{z \in \mathbb{C}: |z| = \eta\}$, is a smooth fibration isomorphic to φ_ε by an isomorphism which preserves the arguments.

We call the fibrations of Theorem 1.1 and 1.2 the Milnor's fibrations of F at 0.

Corollary 1.3 The fibers of φ_ε have the homotopy type of a n-dimensional CW-complex.

Theorem 1.4 Let $V_o = \{f = 0\}$. For $\varepsilon > 0$ small enough, S_ε cuts the smooth part of the algebraic set V_o transversally. If 0 is an isolated critical point of f, then the pairs $(S_\varepsilon, S_\varepsilon \cap V_o)$ for any ε small enough are diffeomorphic, and $(B_\varepsilon, B_\varepsilon \cap V_o)$ is homeomorphic to $(B_\varepsilon, C(S_\varepsilon \cap V_o))$, where $C(S_\varepsilon \cap V_o)$ is the cone which is the union of real line segments joining 0 and points of $S_\varepsilon \cap V_o$.

Theorem 1.4 says that, when 0 is an isolated critical point of f, for $\varepsilon > 0$ small enough, the topology of the pair $(B_\varepsilon, B_\varepsilon \cap V_o)$ does not depend on ε. Then if g is another analytic function defined in a neighborhood of 0, having an isolated critical point at 0, we say that the hypersurface V_o and V'_o defined by $\{f = 0\}$ and $\{g = 0\}$ have the same topological type at 0 if for $\varepsilon > 0$ small enough there is a homeomorphism $(B_\varepsilon, B_\varepsilon \cap V_o) \xrightarrow{\simeq} (B_\varepsilon, B_\varepsilon \cap V'_o)$.

From [16] and [17] we have

Theorem 1.5 If 0 is an isolated critical point of f, for $\varepsilon > 0$ small enough, the fibers of φ_ε have the homotopy type of a bouquet of μ spheres of dimension n with

$$\mu = \dim_{\mathbb{C}} \left(\mathbb{C}\{z_o,\ldots,z_n\}/(\frac{\partial f}{\partial z_o},\ldots,\frac{\partial f}{\partial z_n}) \right)$$

(A bouquet of spheres is the topological space which is a union of spheres having a single point in common.)

We call the number μ of spheres the Milnor number of the critical point 0 of f or the number of vanishing cycles of f at 0. Actually we have

Proposition 1.6 The germ of the morphism $\psi: (\mathbb{C}^{n+1},0) \to (\mathbb{C}^{n+1},0)$ where components are the partial derivatives of f is an analytic cover of degree μ. On the other hand, for $\varepsilon > 0$ small enough, the mapping $\psi_\varepsilon: S_\varepsilon \to S^{2n+1}$ defined by

$$\psi_\varepsilon(z) = \left(\frac{\partial f}{\partial z_0}(z),\ldots,\frac{\partial f}{\partial z_n}(z)\right)\Big/\sqrt{\sum_{i=0}^{n}\left|\frac{\partial f}{\partial z_i}(z)\right|^2}$$

has degree μ.

In [14], Lê Dũng Tráng and C. P. Ramanujan proved the following beautiful theorem.

Theorem 1.7 Let $f(t,z)$ be a polynomial in $z = (z_0,\ldots,z_n)$ with coefficients which are smooth complex valued functions of $t \in I = [0,1]$ such that $F(t,0) = 0$ and such that for each $t \in I$, the polynomials $\frac{\partial F}{\partial z_i}(t,z)$ in z have an isolated zero at 0. Assume moreover that the integer

$$\mu_t = \dim_{\mathbb{C}} \mathbb{C}\{z\}\Big/\left(\frac{\partial F}{\partial z_0}(t,z),\ldots,\frac{\partial F}{\partial z_n}(t,z)\right)$$

is independent of t. Then the monodromy fibrations of the singularities of $F(0,z) = 0$ and $F(1,z) = 0$ at 0 are of the same fiber homotopy. If further $n \neq 2$, these fibrations are even differentiably isomorphic and the topological types of the singularities are the same.

Finally recall a well known result of P. Samuel (cf. [20]) (algebrization of isolated singularities).

Theorem 1.8 Let $f: U \subset \mathbb{C}^{n+1} \to \mathbb{C}$ be an analytic function on a neighborhood U of 0 in $\mathbb{C}^{n+1}$. Suppose $f(0) = 0$ and 0 is an isolated critical point. Then there exists a polynomial $f_o: \mathbb{C}^{n+1} \to \mathbb{C}$ with an isolated critical point at 0 and an analytic isomorphism of a neighorhood U_1 of 0 onto a neighborhood U_2 of 0 which sends the points of $f = 0$ on points of $f_o = 0$.

Let L_n be the set of germs at $0 \in \mathbb{C}^n$ of biholomorphisms $\phi: (\mathbb{C}^n,0) \to (\mathbb{C}^n,0)$. Recall that germs $f,g: (\mathbb{C}^n,0) \to (\mathbb{C},0)$ are called equivalent $f \sim g$ if there exists a $\phi \in L_n$ such that $f = g \cdot \phi$. In the following, we shall classify singularities according to Milnor number. A list of equivalence classes with Milnor number ≤ 10 will be presented.

Theorem 1.9 (Classification Theorem)

For $f: (\mathbb{C}^n,0) \to (\mathbb{C},0)$ with isolated singularity at origin we have,

either: $f \sim g + Q$ where g is a germ of one of the polynomials in the list on the next page, and $Q = z_{r+1}^2 + \ldots + z_n^2$

or: Milnor number $\mu(f) > 10$

Name of the singularity	Equation	Restriction	Milnor Number
A_k	z_1^{k+1}	$1 \leq k \leq 10$	k
D_k	$z_1^2 z_2 + z_2^{k-1}$	$4 \leq k \leq 10$	k
E_6	$z_1^3 + z_2^4$		6
E_7	$z_1^3 + z_1 z_2^3$		7
E_8	$z_1^3 + z_2^5$		8
J_{10}	$z_1^3 + A z_1 z_2^4 + B z_2^6$	$4A^3 + 27B^2 \neq 0$	10

X_9	$z_1^4+tz_1^2z_2^2+z_2^4$	$t^2\neq 4$	9
X_{10}	$z_1^4+z_1^2z_2^2+Az_2^5$	$A\neq 0$	10
P_8	$z_1^3+z_2^2z_3+Az_1z_3^2+Bz_3^3$	$4A^3+27B^2\neq 0$	8
P_9	$z_1z_2z_3+z_1^3+z_2^3+Az_3^4$	$A\neq 0$	9
P_{10}	$z_1z_2z_3+z_1^3+z_2^3+Az_3^5$	$A\neq 0$	10
Q_{10}	$z_1^3+z_2^2z_3+Az_1z_3^3+z_3^4$		10
R_{10}	$z_1^3+z_1z_2z_3+z_2^4+Az_3^4$	$A\neq 0$	10

The proof of this classification theorem can be found in [22]. Since the proof is quite long, we do not include it here. More recently Arnold [1] proved that he had the classification of all singularities with Milnor number $\mu \leq 16$.

§ 2. Milnor number and invariants of strongly pseudoconvex manifolds

A complex manifold of dimension n is a strongly pseudoconvex manifold if there exists a C^∞ real-valued function $\varphi: M \to \mathbb{R}$ such that

(i) For $c \in \mathbb{R}_+$, $\{x \in M: \varphi(x) \leq c\}$ is compact

(ii) The hermitian quadratic form $\Sigma \frac{\partial^2 \varphi}{\partial z_i \partial \bar{z}_j}(z)$ is positive definite outside a compact subset B of M.

In [4], Grauert generalized Cartan's Theorem B to his famous finiteness theorem.

Theorem 2.1 (Grauert) Let M be a strongly pseudoconvex manifold of dimension n. Let F be a coherent analytic sheaf on M. Then $H^p(M,F)$ is a finite dimensional complex vector space, for $p > 0$.

Later, Rossi [19] made the following very nice observation. He proved that a strongly pseudoconvex manifold is a proper modification of a Stein analytic space with isolated singular points. Conversely by Hironaka's resolution of singularities [8], there is always a proper modification of some neighborhood of an isolated singularity which is a strongly pseudoconvex manifold. This fact allows the possibility of studying isolated singular points by pseudoconvexity methods, and vice-versa. Our point of view in this section is to relate the numerical invariants of a strongly pseudoconvex manifold with the numerical invariants of the isolated points of the corresponding Stein analytic space. In [29], the second author proved that $\dim H^i(M;\mathcal{O})$, $1 \leq i \leq n-1$, - the most natural numerical analytic invariants for strongly pseudoconvex manifolds M of dimension n - can be expressed in terms of the local data of the isolated singularities explicitly. Recently, assuming $n = 2$ and the fact that the isolated singularities are hypersurface singularities, Laufer [11] was able to express the Milnor numbers of the isolated singularities in terms of the numerical invariants of M. More specifically, he proved the following theorem.

Theorem 2.2 (Laufer) Let $f(x,y,z)$ be holomorphic in N, a Stein neighborhood of $(0,0,0)$ with $f(0,0,0) = 0$. Let $V = \{(x,y,z) \in N: f(x,y,z) = 0\}$ have $(0,0,0)$ as its only singular point. Let μ be the Milnor number of $(0,0,0)$. Let $\pi: M \to V$ be a resolution of V. Let $A = \pi^{-1}(0,0,0)$. Let $\chi_T(A)$ be the topological Euler characteristic of A. Let K be the canonical divisor on M. Then

$$1 + \mu = \chi_T(A) + K\cdot K + 12 \dim H^1(M,\mathcal{O})$$

It is natural to ask for a formula, valid in higher dimensions, which relates the Milnor number associated to the hypersurface singularity to the numerical invariants of the corresponding strongly pseudoconvex manifold. Indeed, we will prove the following.

Theorem A Let M be a strongly pseudoconvex manifold of dimension $n \geq 3$. Suppose the maximal compact analytic subset in M can be blown down to isolated hypersurface singularities $q_1,\ldots,q_m$. Let Ω^p be the sheaf of germs of holomorphic p-forms on M. Let $\chi^p(M) = \sum_{i=1}^{n} (-1)^i \dim H^i(M,\Omega^p)$. Then

$$m + (-1)^n \sum_{i}^{m} \mu_i = \chi_T(A) + \sum_{p=2}^{n-2} (-1)^{p+1}\chi^p(M) + 2 \sum_{p=0}^{1} (-1)^{p+1}\chi^p(M)$$

here μ_i is the Milnor number of q_i and $\sum_{p=2}^{n-2} (-1)^{p+1}\chi^p(M) = 0$ if $n = 3$ by convention.

<u>Theorem B</u> Let $f(x,y,z)$ be holomorphic in N, a Stein neighborhood of $(0,0,0)$ with $f(0,0,0) = 0$. Let $V_o = N \cap f^{-1}(0)$ have the origin as its only singular point. Let μ be the Milnor number of $(0,0,0)$. Let $\pi\colon M \to V_o$ be a resolution of V_o. Let $A = \pi^{-1}(0,0,0)$. Let $x_T(A)$ be the topological Euler characteristic of A. Let K be the canonical divisor on M. Then

$$1 + \mu \geq x_T(A) + 2 \dim H^1(M,\mathcal{O}) - \dim H^1(M,\Omega^1)$$

and

$$10 \dim H^1(M,\mathcal{O}) + \dim H^1(M,\Omega^1) \geq -K\cdot K$$

We will need the following technical lemma later.

<u>Lemma 2.3</u> Let

$$\begin{array}{ccccccccccccc}
0 \longrightarrow & A_1 & \xrightarrow{\varphi_1} & A_2' \oplus A_2'' & \xrightarrow{\varphi_2} & A_3 & \xrightarrow{\varphi_3} & A_4 & \xrightarrow{\varphi_4} & A_5' \oplus A_5'' & \xrightarrow{\varphi_5} & A_6 \\
& \uparrow{\scriptstyle \pi_1} & & \uparrow{\scriptstyle \pi_2'} \quad \uparrow{\scriptstyle \pi_2''} & & \uparrow{\scriptstyle \pi_3} & & \uparrow{\scriptstyle \pi_4} & & \uparrow{\scriptstyle \pi_5'} \quad \uparrow{\scriptstyle \pi_5''} & & \uparrow{\scriptstyle \pi_6} \\
0 \longrightarrow & B_1 & \xrightarrow{\psi_1} & B_2' \oplus B_2'' & \xrightarrow{\psi_2} & B_3 & \xrightarrow{\psi_3} & B^4 & \xrightarrow{\psi_4} & B_5' \oplus B_5'' & \xrightarrow{\psi_5} & B_6 \\
& & & & & & & & & \| & & \\
& & & & & & & & & 0 & &
\end{array}$$

$$\begin{array}{cccccccc}
\xrightarrow{\varphi_6} & A_7 & \xrightarrow{\varphi_7} & A_8' \oplus A_8'' & \xrightarrow{\varphi_8} & A_9 & \longrightarrow & \cdots\cdots \\
& \uparrow{\scriptstyle \pi_7} & & \uparrow{\scriptstyle \pi_8'} \quad \uparrow{\scriptstyle \pi_8''} & & \uparrow{\scriptstyle \pi_9} & & \\
\xrightarrow{\psi_6} & B_7 & \xrightarrow{\psi_7} & B_8' \oplus B_8'' & \xrightarrow{\psi_8} & B_9 & \longrightarrow & \cdots\cdots \\
& & & \| & & & & \\
& & & 0 & & & &
\end{array}$$

$$\begin{array}{ccccccccc}
\longrightarrow A_{3n-2} & \xrightarrow{\varphi_{3n-2}} & A'_{3n-1} \oplus A''_{3n-1} & \xrightarrow{\varphi_{3n-1}} & A_{3n} & \xrightarrow{\varphi_{3n}} & A_{3n+1} & \xrightarrow{\varphi_{3n+1}} & 0 \\
\uparrow{\scriptstyle \pi_{3n-2}} & & \uparrow{\scriptstyle \pi'_{3n-1}} \quad \uparrow{\scriptstyle \pi''_{3n-1}} & & \uparrow{\scriptstyle \pi_{3n}} & & \uparrow{\scriptstyle \pi_{3n+1}} & & \\
\longrightarrow B_{3n-2} & \xrightarrow{\psi_{3n-2}} & B'_{3n-1} \oplus B''_{3n-1} & \xrightarrow{\psi_{3n-1}} & B_{3n} & \xrightarrow{\psi_{3n}} & B_{3n+1} & \xrightarrow{\psi_{3n+1}} & 0 \\
 & & \| \qquad\qquad & & & & & & \\
 & & 0 \qquad\qquad & & & & & &
\end{array}$$

be a commutative diagram with exact rows. Suppose π''_{3i-1}, π_{3i}, $1 \leq i \leq n$ are isomorphism and all the vector spaces are finite dimensional except possibly A'_2, B'_2, A''_{3i-1}, A_{3i}, B''_{3i-1}, B_{3i}, $1 \leq i \leq n$. Suppose also that $B'_{3i+2} = 0$ for $1 \leq i \leq n-1$. Then

$$\sum_{i=0}^{n} (-1)^i \dim B_{3i+1} + \sum_{i=1}^{n-1} (-1)^i \dim A'_{3i+2} = \sum_{i=0}^{n} (-1)^i \dim A_{3i+1} + \dim \ker \pi'_2 - \dim \operatorname{coker} \pi'_2$$

<u>Proof</u>

$$\sum_{i=0}^{n} (-1)^i \dim B_{3i+1} + \sum_{i=1}^{n-1} (-1)^i \dim A'_{3i+2}$$

$$= \dim \pi_1(B_1) + \dim \ker \pi_1 + \sum_{i=1}^{n} (-1)^i \dim B_{3i}/\psi_{3i-1}(B'_{3i-1} \oplus B''_{3i-1})$$

$$+ \sum_{i=1}^{n-1} (-1)^i \dim \psi_{3i+1}(B_{3i+1}) + \sum_{i=1}^{n-1} (-1)^i \dim A'_{3i+2}$$

$$= \dim \ker \pi_1 - \dim \operatorname{coker} \pi_1 + \dim A_1 + \sum_{i=1}^{n} (-1)^i \dim B_{3i}/\psi_{3i-1}(B'_{3i-1} \oplus B''_{3i-1})$$

$$+ \sum_{i=1}^{n-1} (-1)^i \dim\psi_{3i+1}(B_{3i+1}) + \sum_{i=1}^{n} (-1)^{i-1} (\dim A_{3i}/\varphi_{3i-1}(A'_{3i-1} \oplus A''_{3i-1}$$

$$+ \dim\varphi_{3i+1}(A_{3i+1}) - \dim A_{3i+1}$$

$$+ \sum_{i=0}^{n-1} (-1)^i \dim A'_{3i+2} = \sum_{i=0}^{n} (-1)^i \dim A_{3i+1} + \dim\ker \pi_1 - \dim\operatorname{coker} \pi_1$$

$$+ \sum_{i=1}^{n} (-1)^i \dim B_{3i}/\psi_{3i-1}(B'_{3i-1} \oplus B''_{3i-1}) + \sum_{i=1}^{n-1} (-1)^i \dim\psi_{3i+1}(B_{3i+1})$$

$$+ \sum_{i=1}^{n} (-1)^{i-1} \dim A_{3i}/\varphi_{3i-1}(A'_{3i-1} \oplus A''_{3i-1}) + \sum_{i=1}^{n-1} (-1)^{i-1} \dim\varphi_{3i+1}(A_{3i+1})$$

$$+ \sum_{i=1}^{n-1} (-1)^i \dim A'_{3i+2}$$

$$= \sum_{i=1}^{n} (-1)^i \dim A_{3i+1} + \dim\ker \pi_1 - \dim\operatorname{coker} \pi_1$$

$$+ \sum_{i=1}^{n} (-1)^i \varphi_{3i-1}(A'_{3i-1} \oplus A''_{3i-1})/\pi_{3i} \circ \psi_{3i-1}(B'_{3i-1} \oplus B''_{3i-1})$$

$$+ \sum_{i=1}^{n-1} (-1)^i \dim\psi_{3i+1}(B_{3i+1}) + \sum_{i=1}^{n-1} (-1)^{i-1} \dim\varphi_{3i+1}(A_{3i+1}) + \sum_{i=1}^{n-1} (-1)^i \dim A'_{3i+2} \tag{2.1}$$

Now look at the following commutative diagrams with exact rows.

$$
\begin{array}{ccccccccc}
0 & \longrightarrow & A_1 & \xrightarrow{\varphi_1} & A'_2 \oplus A''_2 & \xrightarrow{\varphi_2} & \varphi_2(A'_2 \oplus A''_2) & \longrightarrow & 0 \\
 & & \pi_1 \uparrow & & \pi'_2 \uparrow \;\; \pi''_2 \uparrow & & \tilde{\pi}_3 \uparrow & & \\
0 & \longrightarrow & B_1 & \xrightarrow{\psi_1} & B'_1 \oplus B''_2 & \xrightarrow{\psi_2} & \psi_2(B'_2 \oplus B''_2) & \longrightarrow & 0
\end{array}
$$

$$
\begin{array}{ccccccccc}
0 & \longrightarrow & \varphi_4(A_4) & \longrightarrow & A'_5 \oplus A''_5 & \longrightarrow & \varphi_5(A'_5 \oplus A''_5) & \longrightarrow & 0 \\
 & & \tilde{\pi}_5 \uparrow & & \pi'_5 \uparrow \;\; \pi''_5 \uparrow & & \tilde{\pi}_6 \uparrow & & \\
0 & \longrightarrow & \psi_4(B_4) & \longrightarrow & B'_5 \oplus B''_5 & \longrightarrow & \psi_5(B'_5 \oplus B''_5) & \longrightarrow & 0 \\
 & & & & \| & & & & \\
 & & & & 0 & & & &
\end{array}
$$

$$\vdots$$

$$
\begin{array}{ccccccccc}
0 & \longrightarrow & \varphi_{3i+1}(A_{3i+1}) & \longrightarrow & A'_{3i+2} \oplus A''_{3i+2} & \longrightarrow & \varphi_{3i+2}(A'_{3i+2} \oplus A''_{3i+2}) & \longrightarrow & 0 \\
 & & \tilde{\pi}_{3i+2} \uparrow & & \pi'_{3i+2} \uparrow \;\; \pi''_{3i+2} \uparrow & & \tilde{\pi}_{3i+3} \uparrow & & \\
0 & \longrightarrow & \psi_{3i+1}(B_{3i+1}) & \longrightarrow & B'_{3i+2} \oplus B''_{3i+2} & \longrightarrow & \psi_{3i+2}(B'_{3i+2} \oplus B''_{3i+2}) & \longrightarrow & 0 \\
 & & & & \| & & & & \\
 & & & & 0 & & & &
\end{array}
$$

$$\vdots$$

$$
\begin{array}{ccccccccc}
0 & \longrightarrow & \varphi_{3n-2}(A_{3n-2}) & \longrightarrow & A'_{3n-1} \oplus A''_{3n-1} & \longrightarrow & \varphi_{3n-1}(A'_{3n-1} \oplus A''_{3n-1}) & \longrightarrow & 0 \\
 & & \tilde{\pi}_{3n-1} \uparrow & & \pi'_{3n-1} \uparrow \;\; \pi''_{3n-1} \uparrow & & \tilde{\pi}_{3n} \uparrow & & \\
0 & \longrightarrow & \psi_{3n-2}(B_{3n-2}) & \longrightarrow & B'_{3n-1} \oplus B''_{3n-1} & \longrightarrow & \psi_{3n-1}(B'_{3n-1} \oplus B''_{3n-1}) & \longrightarrow & 0 \\
 & & & & \| & & & & \\
 & & & & 0 & & & &
\end{array}
$$

$\tilde{\pi}_{3i}$ is injective for $1 \leq i \leq n$. Since $B'_{3i+2} = 0$ for $1 \leq i \leq n-1$ and π''_{3i-1} is an isomorphism for $1 \leq i \leq n$ by snake lemma we have

$$- \dim \ker \pi_1 + \dim \ker \pi'_2 - 0 + \dim \operatorname{coker} \pi_1 - \dim \operatorname{coker} \pi'_2$$

$$\dim\varphi_2(A'_2 \oplus A''_2)/\pi_3 \circ \psi_2(B'_2 \oplus B''_2) = 0 \tag{2.2}$$

$$- (\dim\varphi_4(A_4) - \dim\psi_4(B_4)) + \dim A'_5 - \dim\varphi_5(A'_5 \oplus A''_5)/\pi_6 \circ \psi_5(B'_5 \oplus B''_5) = 0$$
$$\vdots \tag{2.3}$$
$$- (-1)^{i-1}(\dim\varphi_{3i+1}(A_{3i+1}) - \dim\psi_{3i+1}(B_{3i+1}))$$

$$+(-1)^{i-1}\dim A'_{3i+2} - (-1)^{i-1}\dim\varphi_{3i+2}(A'_{3i+2} \oplus A''_{3i+2})/\pi_{3i+3} \circ \psi_{3i+2}(B'_{3i+2} \oplus B''_{3i+2}) = 0$$
$$\vdots \tag{2.4}$$
$$- (-1)^{n-2}(\dim\varphi_{3n-2}(A_{n-2}) - \dim\psi_{3n-2}(B_{3n-2}))$$

$$+ (-1)^{n-2}\dim A'_{3n-1} - (-1)^{n-2}\dim\varphi_{3n-1}(A'_{3n-1} \oplus A''_{3n-1})/\pi_{3n} \circ \psi_{3n-1}(B'_{3n-1} \oplus B''_{3n-1}) = 0 \tag{2.5}$$

Summing (2.1) - (2.5), we get Lemma 2.3. Q.E.D.

Proof of the Theorem A The maximal compact analytic subset A in M has a finite number of connected components $A_\alpha (\alpha = 1,\ldots,m)$. A can be blown down to get a Stein analytic space V_o with isolated hypersurface singularities $q_1,\ldots,q_m$, i.e., $\pi\colon M \to V_o$ is a desingularization of V_o. For $1 \leq \alpha \leq m$ let U_α be a Stein neighborhood of q_α such that $\pi^{-1}(U_\alpha)$ is a holomorphically convex neighborhood of A_α. Then the restriction mapping

$$\gamma\colon H^i(M,\mathcal{J}) \to H^i(U_\pi^{-1}(U_\alpha),\mathcal{J})$$

$$= \oplus\, H^i(\pi^{-1}(U_\alpha),\mathcal{J})$$

where $\mathcal{J}$ is a coherent analytic sheaf is an isomorphism for $i \geq 1$, by Lemma 3.1 of [12]. In order to prove the Theorem A, it suffices to prove the following theorem.

<u>Theorem 2.4</u> Let $f(z_0,z_1,\ldots,z_n)$ be holomorphic in N, a Stein neighborhood of $(0,0,\ldots,0)$ in C^{n+1} with $f(0,0,\ldots,0) = 0$. Let $V_o = \{(z_0,z_1,\ldots,z_n) \in N\colon f((z_0,z_1,\ldots,z_n) = 0\}$ have $(0,0,\ldots,0)$ as its only singular point. Let $\pi\colon M \to V_o$ be a resolution of V_o. Let μ be the Milnor number of $(0,0,\ldots,0)$ and $A = \pi^{-1}(0,0,\ldots,0)$. Suppose $n \geq 3$, then

$$1 + (-1)^n\mu = \chi_T(A) + \sum_{p=2}^{n-2} (-1)^{p+1}\chi^p(M) + 2\sum_{p=0}^{1} (-1)^{p+1}\chi^p(M)$$

where $\sum_{p=2}^{n-2} (-1)^{p+1}\chi^p(M) = 0$ if $n = 3$ by convention.

<u>Proof of Theorem 2.4</u> By Theorem 1.8 any holomorphic function which agrees with f to sufficiently high order, defines a holomorphically equivalent singularity at $(0,0,\ldots,0)$, ([13];[5]). So we may take f to be a polynomial. Compactify C^{n+1} to $\mathbb{P}^{n+1}$. Let $\overline{V}_t$ be the closure in $\mathbb{P}^{n+1}$ of $V_t = \{(z_0,z_1,\ldots,z_n) \in C^{n+1} : f(z_0,z_1,\ldots,z_n) = t\}$. By adding a suitably general high order homogeneous term of degree e to the polynomial f, we may additionally assume that $\overline{V}_o$ has $(0,0,\ldots,0) \in C^{n+1}$ as its only singularity and that $\overline{V}_t$ is non-singular for small $t \neq 0$. We may also assume that the highest order terms of f define, in homogeneous coordinates, a non-singular hypersurface of order e in $\mathbb{P}^{n+1}-C^{n+1}$. $\overline{V}_t$ is then

necessarily irreducible for all small t .

Let $\overline{M}$ be the resolution of $\overline{V}$ which has M as an open subset. We prove that for $0 \leq p \leq n-2$,

$$\sum_{i=0}^{n}(-1)^i \dim H^i(\overline{M}, \Omega^p_{\overline{M}}) = \sum_{i=0}^{n}(-1)^i \dim H^i(\overline{V}_o, \Omega^p_{\overline{V}_o}) + \sum_{i=1}^{n-1}(-1)^i \dim H^i(M, \Omega^p_M) ,$$

i.e.,

$$\chi^p(\overline{M}) = \chi^p(\overline{V}_o) + \chi^p(M) , \quad 0 \leq p \leq n-2 \tag{2.6}$$

as follows: by the Mayer-Vietoris sequence [0] the rows of the following commutative diagram are exact.[1]

$$\begin{array}{l}
0 \to H^0(\overline{M}, \Omega^p_{\overline{M}}) \to H^0(M, \Omega^p_M) \oplus H^0(\overline{M}|A, \Omega^p_{\overline{M}}|_A) \to H^0(M|A, \Omega^p_{M|A}) \qquad (2.7) \\
\qquad \uparrow\pi_1 \qquad\qquad \uparrow\pi_2' \qquad\qquad \uparrow\pi_2'' \qquad\qquad \uparrow\pi_3 \\
0 \to H^0(\overline{V}, \Omega^p_{\overline{V}}) \to H^0(V, \Omega^p_V) \oplus H^0(\overline{V}|\{0\}, \Omega^p_{\overline{V}}|_{\{0\}}) \to H^0(V|\{0\}, \Omega^p_{V|\{0\}}) \\
\to H^1(\overline{M}, \Omega^p_{\overline{M}}) \to H^1(M, \Omega^p_M) \oplus H^1(\overline{M}|A, \Omega^p_{\overline{M}}|_A) \to H^1(M|A, \Omega^p_{M|A}) \to \\
\qquad \uparrow\pi_4 \qquad\qquad \uparrow\pi_5' \qquad\qquad \uparrow\pi_5'' \qquad\qquad \uparrow\pi_6 \\
\to H^1(\overline{V}, \Omega^p_{\overline{V}}) \to H^1(V, \Omega^p_V) \oplus H^1(\overline{V}|\{0\}, \Omega^p_{\overline{V}}|_{\{0\}}) \to H^1(V|\{0\}, \Omega^p_{V|\{0\}}) \to \\
\ldots \to H^{n-1}(\overline{M}, \Omega^p_{\overline{M}}) \to H^{n-1}(M, \Omega^p_M) \oplus H^{n-1}(\overline{M}|A, \Omega^p_{\overline{M}}|_A) \to H^{n-1}(M|A, \Omega^p_{M|A}) \\
\qquad \uparrow\pi_{3n-2} \qquad\qquad \uparrow\pi_{3n-1}' \qquad\qquad \uparrow\pi_{3n-1}'' \qquad\qquad \uparrow\pi_{3n} \\
\ldots \to H^{n-1}(\overline{V}, \Omega^p_{\overline{V}}) \to H^{n-1}(V, \Omega^p_V) \oplus H^{n-1}(\overline{V}|\{0\}, \Omega^p_{\overline{V}}|_{\{0\}}) \to H^{n-1}(V|\{0\}, \Omega^p_{V|\{0\}})
\end{array}$$

---

[1]In this diagram (2.7) we have omitted the subscript zero on the V's for simplicity of notation.

$$\to H^n(\overline{M}, \Omega^p_{\overline{M}}) \to 0$$

$$\uparrow \pi_{3n+1}$$

$$\to H^n(\overline{V}, \Omega^p_{\overline{V}}) \to 0$$

The higher terms in (2.7) are zero by [23]. In (2.7), π_2', π_{3i-1}'', π_{3i}, $1 \leq i \leq n$ are isomorphisms. The fact that π_2' is an isomorphism will be proved later (Theorem (3.6)). Since V is Stein, $H^i(V,\Omega^p_V) = 0$ for all $i \geq 1$. All the vector spaces are finite dimensional except possibly $H^0(M,\Omega^p_M)$, $H^0(V,\Omega^p_V)$, $H^i(\overline{M}|A,\Omega^p_{\overline{M}|A})$, $H^i(M|A,\Omega^p_{M|A})$, $H^i(\overline{V}|\{0\}, \Omega^p_{\overline{V}|\{0\}})$, $H^i(V|\{0\},\Omega^p_{V|\{0\}})$, $0 \leq i \leq n-1$. (2.6) now follows from Lemma 2.3.

In a similar manner, using a tubular neighborhood of A rather than M in (2.7), one sees that

$$\chi_T(\overline{M}) = \chi_T(\overline{V}_0) + \chi_T(A) - 1 \tag{2.8}$$

Recall from Theorem 1.2 that the intersection of V_t with the open ε-ball is diffeomorphic with the fiber F_0. So the manifold-with-boundary $V_t \cap B_\varepsilon$ is connected, with n^{th} Betti number equal to μ, and with Euler number

$$\chi_T(V_t \cap B_\varepsilon) = 1 + (-1)^n \mu .$$

Since the two manifolds $V_t \cap B_\varepsilon$ and $\overline{V}_t$ Int B_ε have union $\overline{V}_t$ and intersection K_t, we have the Euler number of $\overline{V}_t$

$$\chi_T(\overline{V}_t) = \chi(V_t \cap B_\varepsilon) + \chi(\overline{V}_t - \text{Int } B_\varepsilon) - \chi(K_t)$$

$$= 1+(-1)^n\mu + \chi(\overline{V}_0 - \text{Int } B_\varepsilon) - \chi(V_0 \cap S_\varepsilon)$$

by the differentiable triviality of the family $\{V_t\}$ away from $(0,0,\dots,0) \in \mathbb{C}^{n+1}$. Hence

$$\chi_T(\overline{V}_t) = 1 + (-1)^n\mu + \chi(\pi^{-1}(\overline{V}_o \setminus \mathrm{Int}\ B_\varepsilon)) + \chi(\pi^{-1}(V_o \cap B_\varepsilon))$$

$$- \chi(V_o \cap S_\varepsilon) - \chi(\pi^{-1}(V_o \cap B_\varepsilon))$$

$$= 1 + (-1)^n\mu + \chi_T(\overline{M}) - \chi_T(A)$$

since $\pi^{-1}(V_o \cap B_\varepsilon)$ contracts to A. Thus we have

$$1 + (-1)^n\mu = \chi_T(\overline{V}_t) - \chi_T(\overline{M}) + \chi_T(A) \quad . \tag{2.9}$$

Let $b_i(\overline{M})$ be the i-th Betti number of $\overline{M}$ and $h^{q;p}(\overline{M}) = \dim H^p(\overline{M},\Omega^p_{\overline{M}})$. Then

$$\chi_T(\overline{M}) = \sum_{i=0}^{2n} (-1)^i b_i(\overline{M})$$

$$= \sum_{i=0}^{2n} (-1)^i \sum_{p+q=i} h^{p,q}(\overline{M}) \quad \text{(by Hodge decomposition)}$$

$$= \sum_{p=0}^{n} (-1)^p \sum_{q=0}^{n} (-1)^q\ h^{p,q}(\overline{M})$$

$$= \sum_{p=0}^{n} (-1)^p\ \chi^p(\overline{M}) \quad .$$

A similar formula is valid for $\overline{V}_t$. Hence, from (2.9) we have:

$$1 + (-1)^n\mu = \sum_{p=0}^{n} (-1)^p\ \chi^p(\overline{V}_t) - \sum_{p=0}^{n} (-1)^p\ \chi^p(\overline{M}) + \chi_T(A) \quad . \tag{2.10}$$

Let $\overline{V}$ be the hypersurface defined by

$$z_{n+1}^{e}\, f(\frac{z_o}{z_{n+1}}, \frac{z_1}{z_{n+1}}, \ldots, \frac{z_{n-o}}{z_{2n+1}}) - t\, z_{n+1}^{e} = 0$$

in $\mathbb{P}^{n+1} \times D_\varepsilon$ where D_ε is a disk of radious ε in $\mathbb{C}$ and t is the coordinate on D_ε. We consider

$$\Omega^p_{\overline{V}/D_\varepsilon} = \Omega^p_{\overline{V}}/dt \wedge \Omega^{p-1}_{\overline{V}}, \qquad 0 \leq p \leq n$$

(the sheaf of relative p-forms of $\rho: \overline{V} \to D_\varepsilon$, where ρ is the natural projection). We will show in (3.4) that $\Omega^p_{\overline{V}/D_\varepsilon}$ is t-torsion free, i.e. it is <u>flat</u> over D_ε for $p \leq \Omega$. The analytic restriction of $\Omega^p_{\overline{V}/D_\varepsilon}$ on the fibre $\overline{V}_t$ gives the sheaf of holomorphic forms $\Omega^p_{\overline{V}_t}$ of the fibre. Therefore a fundamental result of Grothendieck implies that the Euler-Poincare characteristic $\chi^p(\overline{V}_t) = \sum_{q=0}^{n} (-1)^q h^{p,q}(\overline{V}_t)$ of $\Omega^p_{\overline{V}_t}$ is equal to the Euler-Poincare characteristic

$$\chi^p(\overline{V}_o) = \sum_{q=0}^{n} (-1)^q h^{p,q}(\overline{V}_o) \quad \text{of } \Omega^p_{\overline{V}_o} \quad \text{for } 0 \leq p \leq n \qquad ([6], 7.9.8)\ .$$

From (2.10) we have

$$1 + (-1)^n \mu = \chi_T(A) + \sum_{q=0}^{n} (-1)^q (\chi^q(\overline{V}_o) - \chi^q(\overline{M})) \tag{2.11}$$

$$= \chi_T(A) + \sum_{q=0}^{n-2} (-1)^q (\chi^q(\overline{V}_o) - \chi^q(\overline{M}))$$

$$+ (-1)^n(-1)^n(\chi^o(\bar{V}_o) - \chi^o(\bar{M})) + (-1)^{n-1}(-1)^n(\chi^1(\bar{V}_o) - \chi^1(\bar{M}))$$

$$= \chi_T(A) + \sum_{q=2}^{n-2} (-1)^q(\chi^q(\bar{V}_o) - \chi^q(\bar{M}) + 2(\chi^o(\bar{V}_o) - \chi^o(\bar{M}))$$

$$- 2(\chi^1(\bar{V}_o) - \chi^1(\bar{M}))$$

$$= \chi_T(A) + \sum_{q=2}^{n-2} (-1)^{q+1}\chi^q(M) + 2 \sum_{q=2}^{1} (-1)^{q+1}\chi^q(M)$$

by (2.6) . This proves Theorem A .

Proof of Theorem B

We will use the same notations as in the proof of Theorem A . Set $n = 2$ in (2.11) . We have

$$1 + \mu = \chi_T(A) + \sum_{p=0}^{2} (-1)^p\chi^p(\bar{V}_t) - \sum_{p=0}^{2} (-1)^p\chi^p(\bar{M})$$

$$= \chi_T(A) + 2(\chi^o(\bar{V}_t) - \chi^o(\bar{M})) - (\chi^1(\bar{V}_t) - \chi^1(\bar{M}))$$

$$= \chi_T(A) + 2(\chi^o(\bar{V}_o) - \chi^o(\bar{M})) - (\chi^1(\bar{V}_o) - \chi^1(\bar{M})) \quad .$$

By (2.6) , $\chi^o(\bar{V}_o) - \chi^o(\bar{M}) = - \chi^o(M)$. We will prove later in § 3 that

$$\pi_2^!: \quad H^o(V,\Omega_V^1) \to H^o(M,\Omega_M^1)$$

is injective (see (3.6)). Suppose these for a moment. It follows from (2.7) and Lemma 2.3 that

$$\chi^1(\bar{M}) \geq \chi^1(\bar{V}_o) + \chi^1(M) \ .$$

Hence we have

$$1 + \mu \geq \chi_T(A) - 2\chi^o(M) + \chi^1(M)$$

$$= \chi_T(A) + 2 \dim H^1(M,\mathcal{O}) - \dim H^1(M,\Omega^1) .$$

Combine this with Theorem 2.2 and we get

$$10 \dim H^1(M,\mathcal{O}) + \dim H^1(M,\Omega^1) \geq -K\cdot K .$$

Q.E.D.

Remark: Theorem B is essentially an estimate of $\dim H^1(M,\Omega^1)$. In [31] the second author gives a precise formula for $\dim H^1(M,\Omega^1) + \dim \Gamma(M-A,\Omega^1)/\Gamma(M,\Omega^1)$ in terms of μ, K^2, $\chi_T(A)$ and the number of moduli of the singularity.

§ 3.

In this section we give some results which were cited in the proofs of the theorems above, and which are also of general interest. We carry out the discussion here in the algebraic context. Since the spaces in our complex-analytic application are assumed to be algebrizable (see the beginning of Proof of Theorem A), the results will be immediately applicable to the corresponding complex-analytic sheaves, by means of the usual sheaf-theoretic remarks.

We need to recall a few preliminaries related to the concept of depth; for more details the reader is referred, for example, to [5].

Suppose Y is a closed subspace of V defined by a sheaf of ideals I of $\mathcal{O}_V$. Let $\mathfrak{F}$ be a sheaf on V, $V' = V - Y$. We then have a long exact sequence for local cohomology ("cohomology with supports"):

$$\ldots \to H^i_Y(V,\mathfrak{F}) \to H^i(V,\mathfrak{F}) \to H^i(V',\mathfrak{F}) \to \ldots \quad .$$

If V is affine, and $\mathfrak{F}$ is coherent, we get:

$$0 \to H^0_Y(V,\mathfrak{F}) \to H^0(V,\mathfrak{F}) \to H^0(V',\mathfrak{F}) \to H^1_Y(V,\mathfrak{F}) \to 0 \qquad (3.0.1)$$

$$H^i(V',\mathfrak{F}) \cong H^{i+1}_Y(V,\mathfrak{F}) \quad \text{for } i \geq 1 \ . \qquad (3.0.2)$$

Moreover, on any[(1)] variety V, for coherent we have ([5], 2.8):

$$H^\ell_Y(V,\mathfrak{F}) = \varinjlim_{\nu} \operatorname{Ext}^\ell_{\mathcal{O}_V}(\mathcal{O}_V/I^\nu,\mathfrak{F}) \quad .$$

---

(1) locally noetherian

We will be interested in the case where Y is a point P, so that $I = M_{V,P}$. We recall that $\mathrm{depth}_P(\mathfrak{F})$ is the maximum number of elements in any $\mathfrak{F}$-regular sequence contained in $M_{V,P}$, i.e., a sequence $f_1,\ldots,f_n \in M_{V,P}$ such that f_{i+1} is not a zero divisor in $\mathfrak{F}/(f_1,\ldots,f_i)\mathfrak{F}$. V is Cohen-Macaulay if $\mathrm{depth}_P(\mathcal{O}_V) = \dim_P(\mathcal{O}_V)$ at every point P. We will need

$$\mathrm{depth}_P(\mathfrak{F}) \geq d \Longleftrightarrow \mathrm{Ext}^{\ell}_{\mathcal{O}_V}(\mathcal{O}_V/I,\mathfrak{F}) = 0 \qquad ([5],\ 3.7) \qquad (3.0.4)$$

$$\text{for } \ell \leq d-1 ,$$

where I is any ideal such that $\mathcal{O}_V/I$ is supported at P.

<u>(3.1) Proposition</u> If V is a Cohen-Macaulay affine variety of dimension d, $V' = V-P$ for a point P, then $H^i(V',\mathcal{O}_V) = 0$ for $1 \leq i \leq d-2$.

<u>Proof</u> In view of (3.0.4) (for $\mathfrak{F} = \mathcal{O}_V$), all of the Exts on the right of (3.0.3) vanish. Hence $H^{\ell}_P(V,\mathcal{O}_V) = 0$ for $1 \leq \ell \leq d-1$, and we apply (3.0.2) to get the result.

<u>(3.2) Remark</u> Every smooth variety is Cohen-Macaulay; every relative complete intersection over a Cohen-Macaulay variety is Cohen-Macaulay.

<u>(3.3) Proposition</u> V, P, V' as above with V affine and $\mathfrak{F}$ coherent. The natural map

$$H^0(V,\mathfrak{F}) \to H^0(V',\mathfrak{F})$$

is injective if $\mathrm{depth}_P(\mathfrak{F}) \geq 1$, surjective if $\mathrm{depth}_P(\mathfrak{F}) \geq 2$.

<u>Proof</u> From (3.0.1) we see that the map is injective (resp. surjective) if $H^0_P(V,\mathfrak{F}) = 0$ (resp. $H^1_P(V,\mathfrak{F}) = 0$). The result then follows from

(3.0.3) and (3.0.4).

We can now prove the first of the two main results of this section:

(3.4) Theorem Let T denote the spectrum of a discrete valuation ring with parameter t. Let $V \hookrightarrow \mathbb{A}_T^{n+1}$ be the hypersurface defined by $F = 0$. Assume that $V \to T$ is flat,[2] and smooth outside the origin 0 of the special fibre and $n \geq 1$. Then $\Omega^p_{V/T}$ (the sheaf of relative p-forms of $V \to T$) is flat over T for $p \leq n$ = rel. dim. (V/T).

Proof Let $\Omega = \Omega_{\mathbb{A}_T^{n+1}/T}$, $\bar{\Omega} = \Omega|V = \Omega \otimes_{\mathcal{O}_{\mathbb{A}_T^{n+1}}} \mathcal{O}_V$. Note that for all p, $\Omega^p_{V/T} = \bar{\Omega}^p/dF \wedge \bar{\Omega}^{p-1}$. We want to show that $\Omega^p_{V/T}$ has no t-torsion, i.e.,

$$(*) \quad p \leq n \ , \ \omega \in H^0(\bar{\Omega}^{p-1},V), \ t|dF\wedge\omega \Longrightarrow dF\wedge\omega = dF\wedge t\eta \ ,$$

$$\text{for some } \ \eta \in H^0(V,\bar{\Omega}^{p-1}) \ .$$

Let $V' = V - \{0\}$, which is smooth over T. For each $x \in V'$, there is an open set U of $\mathbb{A}_T^{n+1}$ containing x, on which F may be completed to a system of relative differential parameters of $\mathbb{A}_T^{n+1}$ over T. This means that over U the sheaf Ω is a free $\mathcal{O}_{\mathbb{A}_T^{n+1}}$-module with a free basis provided by the differentials of these parameters. Hence, these differentials also give a free $\mathcal{O}_V$-basis for $\bar{\Omega}$ over $V' \cap U$. In terms of this basis, it is evident that if $t|dF\wedge\omega$, then $dF\wedge\omega|U = dF\wedge t\eta$, for some $\eta \in H^0(U,\bar{\Omega}^{p-1})$. Thus, let us cover V' by open sets U_i, on each of which there is an $\eta_i \in H^0(\bar{\Omega}^{p-1},U_i)$ with

---

(2) The flatness just means that F is not constant on the special fibre.

$$dF \wedge \omega|_{U_i} = dF \wedge t\eta_i \quad .$$

Note that if we could find $\eta \in H^0(V',\bar{\Omega}^{p-1})$ with $dF\wedge\omega|_{V'} = dF\wedge t\eta$, we would be done. In fact, V is flat over T of relative dimension at least 1 , so V is a Cohen-Macaulay variety of dimension ≥ 2, and hence $\operatorname{depth}_o(\mathcal{O}_V) \geq 2$. Thus the section η of the locally free sheaf Ω^{p-1} extends over o by (3.3). And for the extended η, it is true that $dF\wedge\omega = dF\wedge t\eta$ everywhere on V , so that (*) is proved.

To get the $\eta \in H^0(V',\bar{\Omega}^{p-1})$, we try to piece together the η_i : observe that on $U_i \cap U_j$, $\eta_i - \eta_j \in \bar{\Omega}^{p-1}$ is annihilated by dF , so since dF is a basis element of $\bar{\Omega}$, we get $\eta_i - \eta_j \in H^0(U_i \cap U_j, dF\wedge\bar{\Omega}^{p-2})$. Thus the $\eta_{ij} = \eta_i - \eta_j$ give a 1-cocycle on V' with values in $dF\wedge\bar{\Omega}^{p-2}$ for the cover $\{U_j\}$. If this cocycle is cohomologous to 0 , we are done. For then, for each i there is an element $\xi_i \in H^0(U_i dF\wedge\Omega^{p-2})$ with $\xi_i - \xi_j|_{U_i\cap U_j} = \eta_i - \eta_j$, so that the sections $\eta_i - \xi_i$ patch together to give $\eta \in H^0(V',\bar{\Omega}^{p-1})$. We have:

$$dF \wedge \eta|_{U_i} = dF \wedge (\eta_i - \xi_i)|_{U_i} = dF \wedge \eta_i|_{U_i}$$

so that $dF\wedge\omega|_{V'} = dF \wedge t\eta$ as desired.

To complete the proof of (3.1) , we are going to show that $H^1(V', dF\wedge\bar{\Omega}^{p-2}) = 0$. First note that we have a resolution of $dF \wedge \bar{\Omega}^{p-2}$ on V',

$$(**) \quad 0 \to \mathcal{O}_{V'} \to \bar{\Omega}' \to \ldots \to \bar{\Omega}^{p-2} \to dF \wedge \bar{\Omega}^{p-2} \to 0$$

which is obtained by wedging with dF at each stage. Now V is a Cohen-Macaulay scheme of dimension $n+1$ (= rel. dim V/T + dim T) , and $V' = V - 0$

Thus it follows from Proposition (3.1) above that $H^1(V',\mathcal{O}_{V'}) = 0$ for $1 \le i \le n-1$, and since the $\bar{\Omega}^j$ are free sheaves on V', it is also true that $H^i(V',\bar{\Omega}^j) = 0$ for $1 \le i \le n-1$ (and any j). From this, we deduce from (**) that

$$H^1(V',dF\wedge\bar{\Omega}^{p-2}) = H^2(V',dF\wedge\bar{\Omega}^{p-3}) = \ldots = H^{n-1}(V',dF\wedge\bar{\Omega}^{p-n}) .$$

Letting $p = n$, we get:

$$H^1(V',dF\wedge\bar{\Omega}^{n-2}) = H^2(V',dF\wedge\bar{\Omega}^{n-3}) = \ldots = H^{n-1}(V',\mathcal{O}_{V'}) = 0 .$$

Hence, as desired, $H^1(V',dF\wedge\bar{\Omega}^{p-2}) = 0$ for $p \le n$, and the Theorem is proved.

<u>(3.5) Corollary</u> Let $\bar{V}$ be a hypersurface in $\mathbb{P}^{n+1} \times D_\varepsilon$, where D_ε is a disk of radius ε with parameter t, and let $\pi\colon \bar{V} \to D_\varepsilon$ be the projection. Let $\bar{V}_t$ denote $\pi^{-1}(t)$. Suppose $\bar{V}_t$ is smooth for $t \neq 0$, and $\bar{V}_o$ has a single isolated singularity at x. Then $\Omega^p_{\bar{V}/D_\varepsilon}$ is flat over D_ε (i.e., has no t-torsion for $0 \le p \le n$).

<u>Proof</u> Let T denote the spectrum of the algebraic local ring at the origin on the $\mathbb{C}$-line, with parameter t; T is the algebraic version of D_ε. Now let $\bar{V}_a$ denote the algebraic variety in $\mathbb{P}^{n+1} \times T$ which is defined by the same polynomial as $\bar{V}$. $\mathcal{O}_{\bar{V}_a} \subset \mathcal{O}_{\bar{V}}$, for all p, $\Omega^p_{\bar{V}/D_\varepsilon} = \Omega^p_{\bar{V}_a/T} \otimes_{\mathcal{O}_{\bar{V}_a}} \mathcal{O}_{\bar{V}}$. Thus it suffices to prove that $\Omega^p_{\bar{V}_a/T}$ has no t-torsion. Since this is obvious away from x, it suffices to verify it on any neighborhood of x. Thus, choose a $\mathbb{C}^{n+1}$ which contains x in

$\mathbb{P}^{n+1} \times \{0\}$, and let $V = \overline{V}_a \cap (\mathbb{C}^{n+1} \times T) = \overline{V}_a \cap \mathbb{A}_T^{n+1}$. This $V \to T$ satisfies the hypotheses of (3.4) , so we are done.

(3.6) Theorem Let V be a variety of dimension $n \geq 2$, with an isolated hypersurface singularity at a point P . Let $\pi\colon M \to V$ be a desingularization which is an isomorphism outside $\pi^{-1}(P)$. Then the map

$$\pi^*\colon H^0(V,\Omega_V^p) \to H^0(M,\Omega_M^p)$$

is an isomorphism for $0 \leq p \leq n-2$, and is injective for $p = n-1$.

Proof Let $E = \pi^{-1}(P)$. Since $\pi\colon M\text{-}E \to V\text{-}P$ is an isomorphism, the only question is about the p-forms on a neighborhood of P in V . Thus we may assume $V\colon f = 0$ in $\mathbb{C}^{n+1}$.

Let $V' = V\text{-}P$. We first note that to prove the surjectivity (resp. injectivity) of π^* it suffices to prove the same for the natural restriction map

$$\pi\colon H^0(V,\Omega_V^p) \to H^0(V',\Omega_V^p) \quad .$$

In fact, suppose r is surjective, let $\omega \in H^0(M,\Omega_M^p)$ and $\omega' = \omega|_{M-E}$. Thus we can view $\omega' \in H^0(V',\Omega_V^p)$, so by hypothesis it extends to $\eta \in H^0(V',\Omega_V^p)$. Then $\pi^*(\eta) - \omega$ is 0 off E , so since Ω_M^p is locally free (M being smooth), $\pi^*(\eta) = \omega$. Hence π^* is surjective. Moreover, suppose r is injective, and let $\eta \in H^0(V,\Omega_V^p)$, with $\pi^*(\eta) = 0$. Then $\pi^*(\eta)|_{M-E} = \eta|_{V'} = 0$, i.e., $r(\eta) = 0$, so $\eta = 0$. Hence π^* is injective.

Now, according to (3.3) above, the injectivity and surjectivity of r follow from Ω^p having sufficient depth at P ; in fact, we want to show that $\operatorname{depth}_P(\Omega_V^p) \geq 2$ if $p \leq n-2$, and is ≥ 1 for $p = n-1$. And we know from (3.0.4) that this will be true in turn if

$$\operatorname{Ext}^{\ell}_{V}(\mathcal{O}_V/J,\Omega^p_V) = 0 \quad \text{for } \ell \leq 1 \quad \text{if} \quad p \leq n-2 \text{ , and} \qquad (3.6.1)$$
$$\text{for } \ell = 0 \quad \text{if} \quad p = n-1$$

(where J is any ideal such that $\mathcal{O}_V/J$ is supported at P).

To study these Exts , let $\overline{\Omega}^p = \Omega^p_{\mathbb{C}^{n+1}}|_V$; $\overline{\Omega}^p$ is a locally free $\mathcal{O}_V$-module. We have the sequence

$$0 \to \mathcal{O}_V \xrightarrow{d_0} \overline{\Omega}^1 \xrightarrow{d_1} \cdots \to \overline{\Omega}^{p-1} \xrightarrow{d_{p-1}} \overline{\Omega}^p \to \Omega^p_V \to 0 \qquad (3.6.2)$$

where all the maps d_i are given by $\wedge df$. We want to see that this sequence is exact. It is clear that $\Omega^p_V = \overline{\Omega}^p/\operatorname{Im}(d_{p-1})$. The question is whether $\ker d_j = \operatorname{Im} d_{j-1}$ for $j = 1,\dots,p-1$. Since f is a local parameter on $\mathbb{C}^{n+1}$ away from P , the sequence is exact on V' . Thus, let $\omega \in \ker d_j$, where ω is a global section in $H^0(V,\overline{\Omega}^j)$. Then $\omega|_{V'} \in \operatorname{Im}(d_{j-1}|_{V'})$, say $\omega|_{V'} = d_{j-1}(\eta')$, $\eta' \in H^0(V',\overline{\Omega}^{j-1})$. Now $\operatorname{depth}_p(\overline{\Omega}^{j-1}) = n$ since $\overline{\Omega}^{j-1}$ is locally free and $\operatorname{depth}_p(\mathcal{O}_V) = n$ (because V is a hypersurface; see remark (3.2)). Hence, since $n \geq 2$, by the surjectivity assertion of (3.3) , η' extends to a section η of $H^0(V,\overline{\Omega}^{j-1})$; $d_{j-1}(\eta)$ agrees with ω away from P , so they actually agree in $\overline{\Omega}^j$ by the injectivity assertion of (3.3) , i.e., $d^{j-1}(\eta) = \omega$.

Thus (3.6.2) is, locally at P , a free $\mathcal{O}_V$ resolution of Ω^p_V . The point is then a resolution of Ω^p_V by sheaves of depth n at P , i.e.,

$$\text{For each } j = 0,\dots,p \quad \mathrm{Ext}^{\ell}_{\mathcal{O}_V}(\mathcal{O}_V/J,\bar{\Omega}^j) = 0 \text{ for } \ell \le n-1 . \tag{3.6.3}$$

Note, moreover, that (3.6.2) is equivalent to the following list of p short exact sequences:

$$\begin{array}{ccccccccc}
0 & \longrightarrow & \Omega_V^{p-1} & \longrightarrow & \bar{\Omega}^p & \longrightarrow & \Omega_V^p & \longrightarrow & 0 \\
0 & \longrightarrow & \Omega_V^{p-2} & \longrightarrow & \bar{\Omega}^{p-1} & \longrightarrow & \Omega_V^{p-1} & \longrightarrow & 0 \\
 & & & & \vdots & & & & \\
0 & \longrightarrow & \Omega_V^{s-1} & \longrightarrow & \bar{\Omega}^s & \longrightarrow & \Omega_V^s & \longrightarrow & 0 \\
 & & & & \vdots & & & & \\
0 & \longrightarrow & \mathcal{O}_V & \longrightarrow & \bar{\Omega}^1 & \longrightarrow & \Omega_V^1 & \longrightarrow & 0 .
\end{array}$$

For each of these short exact sequences - i.e., for $s = 1,\dots,p$ - we get a long exact sequence

$$\dots \to \mathrm{Ext}^{\ell-1}_{\mathcal{O}_V}(\mathcal{O}_V/J,\bar{\Omega}^s) \to \mathrm{Ext}^{\ell-1}_{\mathcal{O}_V}(\mathcal{O}_V/J,\Omega_V^s) \to \mathrm{Ext}^{\ell}_{\mathcal{O}_V}(\mathcal{O}_V/J,\Omega_V^{s-1}) \to \dots \quad .$$

We deduce that if $\ell \le n-1$,

$$\mathrm{Ext}^{\ell-1}_{\mathcal{O}_V}(\mathcal{O}_V/J,\Omega_V^s) \xrightarrow{\sim} \mathrm{Ext}^{\ell}_{\mathcal{O}_V}(\mathcal{O}_V/J,\Omega_V^{s-1}) \quad .$$

Hence, by iterating we see that

$$\mathrm{Ext}^{0}_{\mathcal{O}_V}(\mathcal{O}_V/J,\Omega_V^p) = \mathrm{Ext}^{p}_{\mathcal{O}_V}(\mathcal{O}_V/J,\mathcal{O}_V) \quad \text{if} \quad p \le n-1$$

and

$$\mathrm{Ext}^{1}_{\mathcal{O}_V}(\mathcal{O}_V/J,\Omega_V^p) = \mathrm{Ext}^{p+1}_{\mathcal{O}_V}(\mathcal{O}_V/J,\mathcal{O}_V) \quad \text{if} \quad p \le n-2 \quad .$$

However, by (3.1.3), if $p \leq n-1$, $\mathrm{Ext}^p_{\mathcal{O}_V}(\mathcal{O}_V/J\mathcal{O}_V) = 0$ and if $p \leq n-2$, $\mathrm{Ext}^{p+1}_{\mathcal{O}_V}(\mathcal{O}_V/J\mathcal{O}_V) = 0$. Thus we have verified (3.6.9) and the proof of the theorem is completed.

Bruce Bennett
Department of Mathematics
University of California
Irvine, California

Stephen S.-T. Yau
University of Illinois
at Chicago Circle

Present Address:

Department of Mathematics
Princeton University
Princeton, New Jersey 08544

Research partially supported by N.S.F. Grants and Princeton University

References

0. Andreotti, A. and Grauert, H., Theoremes de finitude pour la cohomologie des espaces complexes, Bull. Soc. Math. France 90 (1962), 193-259.

1. Arnold, V., Local normal forms of functions, Inv. Math. 35, 87-109 (1976).

2. Artin, M., On isolated rational singularities of surfaces, Amer. J. Math. 88, 129-136 (1966).

3. Eells, J. and Lemaire, L., A report on harmonic maps, Bull. London Math. Soc. 10 (1978), 1-68.

4a. Eells, J. and Sampson, J. H., Harmonic mappings of Riemannian manifolds, Amer. J. Math., 86 (1964), 109-160.

4b. Grauert, H., On Levi's Problem and the imbedding of real-analytic manifolds, Ann. Math. (1958), 460-472.

5. Grothendieck, A., Local Cohomology, Lecture Notes in Mathematics, Vol. 41, Springer 1967.

6. Grothendieck, A., Éléments de Géometric Algebrique (III, part 2), Publications Mathématiques de l'I.H.E.S. No. 17, 1963.

7. Hartman, P., On homotopic harmonic maps, Canad. J. Math. 19 (1967), 673-687.

8. Hironaka, H., Resolution of singularities of an algebraic variety over a field of characteristic 0, Ann. Math. 79 (1964), 109-326.

9. Hironaka, H., A fundamental lemma on point modifications, Conference on Complex Analysis, Minneapolis, Springer-Verlag, 1965.

10. Hironaka, H., and Rossi, H., On the equivalence of imbeddings of exceptional complex spaces, Math. Ann. 156 (1964), 313-323.

11. Laufer, H., On μ for surface singularities, Proc. Sym. Pure Math. XXX, Part I, Amer. Math. Soc., Providence, 45-50.

12. Laufer, H., On rational singularities, Amer. J. Math. 94 (1972), 597-608.

13. Laufer, H., On minimally elliptic singularities, Amer. J. Math. 99, 1257-1295 (1977).

14. Le Dung Trang and Ramanujan, C. P., The invariance of Milnor's number implies the invariance of the topological type, Amer. J. Math., Vol. 98, No. 1, (1976), 67-78.

15. Mather, J., (unpublished, handwritten notes, apparently widely distributed, often referred to as Notes on Right-equivalence), 1970.

16. Milnor, J., Singular points of complex hypersurfaces, Ann. of Math. Studies, No. 61, Princeton University Press, Princeton, 1968.

17. Palamodov, V. P., Multiplicity of holomorphic mappings, Funct. Anal. and Appl. (1967), 218-226. = Funktsional'nyi Analizi Ego Prilozheniya 1, #3 (1967), 54-65

18. Riemenschneider, O., Uber die Anwendung algebraischer Methoden in der Deformationstheorie komplexer Räume, Math. Ann. 187 (1970), 40-55.

19. Rossi, H., Strongly pseudoconvex manifolds, Lecture Notes in Mathematics 103, Springer-Verlag, 1969.

20. Samuel, P., Algebricite de certains points singuliers algebroids, J. Math. Pure Appl. 35 (1956), 1-6.

21. Siersma, D., The singularities of C^∞-functions of right-codimension smaller or equal than eight, Indag. Math. 35, No. 1, 31-37, (1973).

22. Siersma, D., Classification and deformation of singularities, (Second printing 1979 by Mathematics Department of the University of Utrecht), 1974.

23. Siu, Y.-T., Analytic sheaf cohomology groups of dimension n of n-dimensional complex spaces, Trans. A.M.S. 143 (1969), 77-94.

24. Siu, Y.-T., The complex analyticity of harmonic maps and strong rigidity of compact Kahler manifolds, Ann. Math. 112 73-111 (1980).

25. Siu, Y.-T., Some remarks on the complex-analyticity of harmonic maps, S.E.A. Bull. Math. 3 (1979), 240-253.

26. Wood, J. C., Harmonic mappings between surfaces, Thesis, Warwick University, (1974).

27. Wood, J. C., Singularities of harmonic maps and applications of Gauss-Bonnet formula, Amer. J. Math. (1978).

28. Yau, Stephen S.-T., Vanishing Theorems for Resolutions of Higher Dimensional Singularities, Hironaka Numbers and Elliptic Singularities in Dimensions > 2 . (Preprint).

29. Yau, Stephen S.-T., The Signature of Milnor Fibres and Duality Theorem for Strongly Pseudoconvex Manifolds, Inv. Math. 46, 81-97, (1978).

30. Yau, Stephen S.-T., Gorenstein singularities with goemetric genus equal to two, Amer. J. Math. 101, 813-854, (1979).

31. Yau, Stephen S.-T., Various numerical invariants for isolated singularities, (preprint).

HARMONIC CURVATURE FOR GRAVITATIONAL AND YANG-MILLS FIELDS

by

Jean-Pierre BOURGUIGNON

Centre de Mathématiques*
Ecole Polytechnique

F - 91128 PALAISEAU CEDEX
(France)

These notes intend to compare some geometric variational problems (e.g. Yang-Mills connections or Einstein metrics) with the harmonic map problem to which we refer in brackets [...]. The notations of the lectures by J. Eells will be used. Emphasis is put on differences in the structures of the various theories and on analogies between the results which have so far been obtained.

No new results are included, but by these notes the author hopes to help newcomers to the field by "revealing" some basic features of the theories commonly burried under technicalities in research papers (although they are the keys to many natural questions).

The style has been kept close to that of the lecture. For detailed proofs or even for full statements, we refer to [2] or [3] Parts II and III. No comment will be made on the physical relevance of the theories discussed, see [3] Part I or [7] for (brief) introductions.

*Laboratoire associé au C.N.R.S. n° 169

Part One. A brief review of Yang-Mills theory

I - The basic set-up

Yang-Mills theory deals with vector bundles over a Riemannian manifold such as $\pi : V \to (M,g)$.

[In harmonic map theory the bundles under consideration are trivial bundles ($\pi: M \times N \to M$) where π is the first projection), but the fibre is not necessarily a linear space.]

We will suppose that the fibres are endowed with some extra structure, so that $\pi : W \to (M,g)$ is in fact an associated bundle for some principal G-bundle $p : P \to M$ where G is a compact Lie group.

For Yang-Mills theory, it turns out that compact Lie groups have to be split into three families :

. the abelian groups $U_1, U_1 \times U_1, \dots, \mathbb{T}^k = \underbrace{U_1 \times \dots \times U_1}_{k \text{ times}}$;

. the small groups SU_2 , U_2 , SU_3 and their products with abelian groups. They are characterized as the compact Lie groups in which the centralizer of any non-central element is an abelian group;

. all the other groups starting with SO_4 .

For Yang-Mills theory the competitors in the variational problem are G-connections on the G-bundle $V \to (M,g)$. (We denote this space by $\mathcal{C}$.)

[In harmonic map theory, the competitors are maps from M to N, which we may like to view as sections of $M \times N \to M$.]

We regard a G-connection as a first order differential operator $\nabla : \Omega^o(M,V) \to \Omega^1(M,V)$ whose principal symbol is the identity (here $\Omega^i(M,V)$ denote the space of V-valued differential i-forms on M). In other words, aside from linearity with respect to scalars, ∇ satisfies

$$\nabla(fs) = f\,\nabla s + df \otimes s$$

for all C^∞ functions f on M and C^∞ sections s of V. What makes ∇ a G-connection is rather technical to be discussed in full generality. When the G-structure is defined by some tensorial data in each fibre (e.g. a fibre metric for G = SO , or a compatible triple of a complex structure, a hermitian metric and a complex volume element for $G = SU$), the tensorial data must be annihilated by (the tensorial extension of) ∇ , i.e. be parallel.

One easily sees that the space of G-connections is an affine space modeled after the vector space $\Omega^1(M,\mathfrak{g}_V)$. Here $\mathfrak{g}_V$ denotes the image of the adjoint bundle $\mathfrak{g}_P = P\times_{ad}\mathfrak{g}$ by the representation $\rho : G \to \xi^*\otimes\xi$ so that $V = P\times_\rho \xi$ (we suppose ρ locally faithful). By construction $\mathfrak{g}_V$ is a sub-Lie algebra bundle of $V^*\otimes V$.

The Yang-Mills energy is taken to be the functional

$$E(\nabla) = \frac{1}{2}\int_M \|R^\nabla\|^2 \, v_g$$

where R^∇ denotes the curvature of the connection ∇ . Recall that R^∇ belongs to $\Omega^2(M,\mathfrak{g}_V)$ and can be defined as $d^\nabla\nabla$ where d^∇ denotes the exterior differential on V-valued forms deduced from ∇ . To compute $\|R^\nabla\|^2$ we use the natural norm defined by g on 2-forms on M and a chosen G-invariant norm on the fibres of $\mathfrak{g}_V$, e.g. the Killing form of G (The coordinate expression of the energy density would then be $g^{ij}\, g^{k\ell}\, R^\nabla_{ik\alpha}{}^{\beta} R^\nabla_{j\ell\beta}{}^{\alpha}$.)

[For a map $\varphi : M \to N$ the energy is given by

$$E(\varphi) = \frac{1}{2}\int_M \|d\varphi\|^2 \, v_g \, .$$

The energy density involves also two metrics : the metric g on M and the metric h on N (a sort of fibre metric). Notice that $d\varphi$ belongs to $\Omega^1(M,\varphi^* TN)$, i.e. is a 1-form with values in a bundle which varies with φ . This point makes the theory at the same time more and less complicated than Yang-Mills theory.]

At this point it may be worth explaining why the degrees of the forms under consideration are different in the two theories.

More generally we look at the jet bundle J^kW of a general bundle $\pi : W \to M$. A (geometric) variational problem can be attached to any energy density e , $e : J^kW \to \mathbb{R}$ by defining the total energy of a C^∞ section s of $\pi : W \to M$ as

$$E(s) = \int_M e(j_k(s))\, d\mu$$

(we also need a density $d\mu$ on M , e.g., the one given by a Riemannian metric).

For harmonic map theory $k = 1$ (a map is a zeroth order object, hence its differential a first order object). For Yang-Mills theory $k=2$ (a connection is already a first order object, hence its curvature is a second order object which turns out to be tensorial and skew because of the symmetry of second derivatives). This difference is reflected in the degree of the forms involved, a 1-form for harmonic map theory, a 2-form for Yang-Mills theory. We will see in the next section how this is reflected in the special dimensions in which the theories are especially interesting.

II - Special dimensions

When M is oriented it is convenient to concentrate the occurence of the metric g into the Hodge map $*$, $* : \Lambda^j T^*M \to \Lambda^{m-j} T^*M$ (here $m=\dim M$). We get

$$E(\nabla) = \frac{1}{2} \int_M |R^\nabla \wedge *R^\nabla|$$

and $|\ |$ refers only to the fibre metric.

[Analogously one gets

$$E(\varphi) = \frac{1}{2} \int_M |d\varphi \wedge *d\varphi| .]$$

It is easy to check (and probably not enough known) that $*$ depends only on the conformal class defined by g on j-forms when $m=2j$. One then has

$$*^2 = (-1)^j .$$

This singles out dimension 4 for Yang-Mills theory as the only dimension where the theory is conformally invariant. For $m=4$, $*$ is an involution of T_xM for each point x ($T_xM = T$ henceforth). One can then decompose Λ^2T into $\Lambda^+T \oplus \Lambda^-T$ where $\Lambda^{\pm}T$ denotes the $\pm$ 1-eigenspaces of $*$, respectively called the spaces of self-dual and anti-self dual forms.

It is important to notice that after identifying Λ^2T with the Lie algebr SO_T, the decomposition $\Lambda^2T = \Lambda^+T \oplus \Lambda^-T$ which is special to $\dim T = 4$ fits with the decomposition of SO_T into its simple ideals isomorphic to SU_2.

[Harmonic map theory is conformally invariant when $m=2$ (recall that m is the dimension of the source space). This is a mere generalization of the conforma invariance of the Dirichlet integral.

The Hodge map defines then a complex structure, which ties together harmonicity and holomorphy (or anti-holomorphy). Notice that to get a decomposition of the bundle of 1-forms one is forced to complexify it.]

It is therefore natural to decompose the curvature into its self-dual and anti self-dual parts,

$$R^{\nabla} = R^{\nabla}_{+} + R^{\nabla}_{-}$$

and to split the energy as $E(\nabla) = E^{+}(\nabla) + E^{-}(\nabla)$ where, for $\varepsilon = \pm$, we set

$$E^{\varepsilon}(\nabla) = \frac{1}{2}\int_M \|R^{\nabla}_{\varepsilon}\|^2 v_g .$$

The crucial point is then the Chern-Gauss-Bonnet theorem which states that

$$E^{+}(\nabla) - E^{-}(\nabla) = 4\pi^2 p_1(V)$$

(where $p_1(V)$ is the first Pontryagin number of V). The functional $\nabla \to E^{+}(\nabla) - E^{-}(\nabla)$ is therefore independent of ∇.

As a trivial corollary we obtain that a connection ∇ whose curvature is ε self-dual (where ε is the sign of $-p_1(V)$) is an absolute minimum of the energy hence a Yang-Mills connection (such connections are called instantons).

On $M = S^4$ and for G a simple group, these solutions have been fully described by Atiyah-Drinfeld-Hitchin-Manin [5]. They used the twistor construction due to Pensore and results in algebraic geometry reducing the problem to some cohomological computations.

An alternate description should be obtainable by pulling back the natural connection on the Hopf bundle over the quaternionic projective space $\mathbb{H}P^n$ by some specific maps from $S^4 = \mathbb{H}P^1$ into $\mathbb{H}P^n$. This would keep results in Yang-Mills theory within the realm of differential geometry to which it really belongs (until analysis has swallowed it !).

This is in complete analogy with the decomposition $E(\varphi) = E'(\varphi) + E''(\varphi)$ of the energy $E(\varphi)$ of a map φ into its holomorphic and antiholomorphic part. This decomposition makes sense for almost complex M and N's.

The difference $E' - E''$ is an invariant of the homotopy class of φ as soon as M is compact and N Kählerian. This dependence on connected components of the space of maps from M to N is due to the fact that $d\varphi$ takes its values in the bundle $\varphi^* TN$ whose topology depends on φ.]

When $G = SO_4$, one can push the decomposition one step further. Indeed since $SO_4 = SU_2 \oplus SU_2$, one can write

$$R^\nabla = R_+^{\nabla+} + R_+^{\nabla-} + R_-^{\nabla+} + R_-^{\nabla-} ,$$

the upper signs are refering to a Hodge map for 2-forms along the fibres.

This decomposition being orthogonal, one can split the energy E into

$$E = E_+^+ + E_+^- + E_-^+ + E_-^- .$$

These functionals are tied together by two constraints following from Chern-Gauss-Bonnet theorems, namely

$$E_+^+(\nabla) + E_+^-(\nabla) - E_-^+(\nabla) - E_-^-(\nabla) = 4\pi^2 p_1(V) ,$$

$$E_+^+(\nabla) - E_+^-(\nabla) - E_-^+(\nabla) + E_-^-(\nabla) = 8\pi^2 \chi(V) .$$

As a result, any two of those functionals determine the two others. Depending on the relative values of $p_1(V)$ and $\chi(V)$, only two components of the energy can vanish at the same time. When this is the case, the energy achieves an absolute minimum which is not necessarily ε self-dual (think of the tangent bundle of S^4 with its standard metric !) but <u>two-fold self-dual</u>. These connections are more general than direct sums of ε self-dual connections.

Back to general G but keeping M oriented 4-dimensional, an object of primary interest is the analogue of the holomorphic quadratic differential $(\varphi^* h)_{2,o}$ introduced in harmonic map theory, namely $[R_+^\nabla , R_-^\nabla]$ which can be defined as follows :

$R_\pm^\nabla$ being an element of $\Omega^\pm(M, \mathfrak{g}_V)$ for each point x of M , $R_+^\nabla(x) \otimes R_-^\nabla(x)$ belongs to $\Lambda^+T^* \otimes \Lambda^-T^* \otimes (\mathfrak{g}_V)_x \otimes (\mathfrak{g}_V)_x$. Using the Lie algebra structure of each fibre of $\mathfrak{g}_V$, we can map $(\mathfrak{g}_V)_x \otimes (\mathfrak{g}_V)_x$ to $(\mathfrak{g}_V)_x$ and therefore get $[R_+^\nabla , R_-^\nabla](x)$

The vector spaces Λ^+T and Λ^-T are both 3-dimensional, so that $\Lambda^+T \otimes \Lambda^-T$ is 9-dimensional. It is an elementary (but generally overlooked) fact that $\Lambda^+T \otimes \Lambda^-T$ is naturally isomorphic as SO_T-module to S_o^2T , the space of traceless symmetric 2-tensors on T . This simple algebraic property turns out to be of overwhelming importance in dealing with stable Yang-Mills fields (see IV).

III - Variational theory

One is looking for extremals of the energy. If ∇^t is a curve of connections issuing from ∇ (e.g., $\nabla^t = \nabla + tA$ where A belongs to $\Omega^1(M,\mathfrak{g}_V)$, then

$$R^{\nabla^t} = R^{\nabla} + t\, d^{\nabla}A + t^2 [A \wedge A]$$

where $[A \wedge A]$ is a $\mathfrak{g}_V$-valued 2-form obtained from the $\mathfrak{g}_V \otimes \mathfrak{g}_V$-valued 2-form $A \wedge A$ by using the Lie algebra structure of $\mathfrak{g}_V$.

The first variational formula is

$$\frac{d}{dt} E(\nabla^t)\Big|_{t=o} = \frac{1}{2}\int_M ((dA, R^{\nabla}))\, v_g .$$

Therefore the Euler-Lagrange equation for the Yang-Mills energy is

$$(d^{\nabla})^* R^{\nabla} = 0$$

where $(d^{\nabla})^*$ denotes the formal adjoint of d^{∇}. (Notice that $(d^{\nabla})^*$ involves the Riemannian metric g on M via the Hodge map $*$ since $(d^{\nabla})^* = -* \circ d^{\nabla} \circ *$.)

Since the second Bianchi identity $d^{\nabla}R^{\nabla} = 0$ always holds here, a critical point ∇ of the Yang-Mills energy, a so-called Yang-Mills connection, is a G-connection with harmonic curvature.

Notice that when G is an abelian Lie group, R^{∇} varies linearly with ∇. The energy is then a quadratic functional of ∇ and the whole theory reduces to ordinary Hodge theory of 2-forms.

[The Euler-Lagrange equation for a map φ reads $d^* d\varphi = 0$ (recall that $d^* d\varphi$ belongs to $\Omega^o(M,\varphi^* TN)$). In this case one also has the identity $d^h d\varphi \equiv 0$ which expresses the symmetry of second derivatives defined by using the pull-back of the Levi-Civita connection on the bundle $\varphi^* TN \to M$. This justifies the critical points of the energy be called harmonic maps.]

As usual the second variational formula will go deeper into the geometry of the situation.

At a Yang-Mills connection ∇, the second variational formula reads

$$\frac{d^2E}{dt^2}(\nabla^t)\Big|_{t=o} = \int_M (((d^{\nabla})^* d^{\nabla}A + 2\sum_{i=1}^{m} [R^{\nabla}_{e_i,\cdot}, A_{e_i}], A))\, v_g$$

where (e_i) is a g-orthonormal basis at a point x. The operator of the second variation which we denote by $\mathcal{S}^\nabla$ is a second order differential operator on $\Omega^1(M,\mathfrak{g}_V)$ as expected, but $\mathcal{S}^\nabla$ is not elliptic.

[The Jacobi operator J_φ acting on $\Omega^1(M,\varphi^* TN)$ is elliptic and generalizes straightforwardly the Jacobi operator along a geodesic].

Let us explain why the operator $\mathcal{S}^\nabla$ cannot be elliptic. On the space of connections $\mathcal{C}$, the group $\mathcal{G}$ of sections of the automorphism bundle G_V, the so-called <u>gauge group</u>, acts as follows : if g belongs to $\Omega^o(M,G_V)$, then

$$\nabla^g = g^{-1} \circ \nabla \circ g \quad .$$

The gauge group $\mathcal{G}$ is infinite dimensional. At a connection ∇, the tangent space to the orbit $\mathcal{G}.\nabla$ is $\operatorname{Im} d^\nabla$. (Indeed if γ is an element of the Lie algebra $\Omega^o(M,\mathfrak{g}_V)$ of $\mathcal{G}$, then

$$\frac{d}{dt}\nabla^{g_t}\Big|_{t=o} = [\nabla,\gamma]$$

$$= d^\nabla\gamma$$

by definition of the natural extension of ∇ to tensor bundles over V.)

The energy is invariant under $\mathcal{G}$ since $R^{\nabla^g} = g^{-1} \circ R^\nabla \circ g$ and since, at each point x of M, $g(x)$ is an orthogonal transformation for the chosen fibre metric.
It then follows that, for all γ in $\Omega^o(M;\mathfrak{g}_V)$,

$$\mathcal{S}^\nabla(d^\nabla\gamma) = 0 \ .$$

Hence the kernel of $\mathcal{S}^\nabla$ is infinite-dimensional and $\mathcal{S}^\nabla$ cannot be elliptic.

As the non ellipticity comes from invariance of E under a group action, it is natural to normalize by working transversally to the orbit directions. This can be done as follows : since d^∇ is a differential operator with injective symbol, one has the following decomposition

$$T_\nabla\mathcal{C} = \Omega^1(M,\mathfrak{g}_V) = \operatorname{Im} d^\nabla \oplus \operatorname{Ker} (d^\nabla)^* \ .$$

One can then normalize the variations A of the connections by supposing that $(d^\nabla)^* A = 0$. On such normalized variations, the second variation operator can be taken to be

$$\tilde{\mathcal{J}}^{\nabla} = \mathcal{J}^{\nabla} + d^{\nabla}(d^{\nabla})^{*}$$

which is nicely elliptic.

[The occurence of this infinite dimensional invariance group should not be confused with the conservation law established by P. Baird and J. Eells involving the so-called stress energy tensor. This conservation law comes from the action of the group of diffeomorphisms of the source space which does not preserve the metric g on M and therefore does not leave the energy invariant. Only the group of isometries of (M,g) or in dimension 2 the group of conformal transformations leaves the energy invariant. Its analog in Yang-Mills theory is the enlarged gauge group, i.e., the group of automorphisms of the bundle covering an isometry of (M,g) or in dimension 4 a conformal transformation.]

Among critical points the local minima are of special interest since they are stable. More generally we call weakly stable a Yang-Mills connection ∇ for which the operator $\tilde{\mathcal{J}}^{\nabla}$ is non-negative.

IV - A few results about Yang-Mills connections

Using special variations of a connection related to the action of the (non-compact part of the) conformal group on the standard sphere S^n, J. Simons proved that a Yang-Mills connection ∇ on any G-bundle over S^n for $n \geq 5$ cannot be stable.

[Using similar arguments, S.L. Xin proved in [9] that the identity of S^n is not a stable harmonic map for $n \geq 3$.]

In [2], a joint paper with H.B. Lawson, we prove that on S^4 any weakly stable Yang-Mills connection ∇ is such that $[R_+^{\nabla}, R_-^{\nabla}] = 0$.

For the "small" non abelian Lie groups SU_2, U_2, SU_3 this is enough to show that ∇ is ε-self-dual. For $G = SO_4$ one can conclude that ∇ is two fold self-dual. For larger groups it is likely that the vanishing of $[R_+^{\nabla}, R_-^{\nabla}]$ implies reducibility or ε-self-duality, but this is still open.

Similar arguments can be carried over to any orientable homogeneous 4-manifold, ([3] Part III), i.e., for $M = \mathbb{C}P^2$, $S^1 \times S^3$, $S^2 \times S^2$, T^4, $T^2 \times S^2$. The only new phenomenon is that, when M has some non trivial 2-cohomology, one may have to twist the bundle V by a line bundle.

[For a harmonic map φ over a Riemann surface, similar reasonings aim at concluding that φ is holomorphic or antiholomorphic when φ is a minimum of the energy.]

So far nobody has found the operator which annihilates $[R_+^\nabla, R_-^\nabla]$ when ∇ is a Yang-Mills connection.

[The 2-form $(\varphi^* h)_{(2,o)}$ is holomorphic as soon as φ is harmonic and M a Riemann surface.]

Part Two. A quick survey of gravitational theory

In gravitational theory we will vary the Riemannian metric g on the manifold M. Therefore the bundle under consideration is the tangent bundle. The main new feature is that the diffeomorphism group of the manifold M acts naturally on tensor fields.

The most commonly used functional is the gravitational energy

$$E(g) = \int_M u_g \, v_g$$

where u_g denotes the scalar curvature of g. The study is of interest only after normalizing suitably the total volume $\int_M v_g$ (we will take $\int_M v_g = 1$).

The critical points of the energy are the Einstein metrics g characteriz by the equation

$$r_g - \frac{1}{n} u_g \, g = 0$$

where r_g denotes the Ricci curvature of the metric g.

Another functional which appears to be of interest since it involves the whole curvature tensor is $\tilde{E}(g) = \frac{1}{2} \int_M \|R^{\nabla^g}\|^2 \, v_g$ (here of course ∇^g denotes the Levi-Civita connection of the metric g).

The Euler-Lagrange equation of the functional $\tilde{E}$ is fourth order in g (this is why physicists did not pay too much attention to it, convinced as they are that the physically relevant equations have to be of second order however compare [8]). An intermediate case is obtained by considering metrics with harmonic curvature i.e., metrics for which

$$(d^{\nabla^g})^* R^{\nabla^g} = 0 .$$

This is a third order system in the metric g. Among these metrics one finds the Einstein metrics and the conformally flat metrics with constant scalar curvature. This harmonicity condition appears in some sense when one uncouples the Riemannian metric g on the base and its Levi-Civita connection D on the tangent bundle. The third order system in g means that D is a Yang-Mills connection while keeping the metric on M fixed.

In [4] A. Derdzinski constructed metrics with harmonic curvature on product manifolds $S^1 \times M'$ for any M' carrying an Einstein metric with positive scalar curvature. The Ricci tensors of those metrics are not parallel (hence the metrics not products of Einstein metrics). The case $S^1 \times S^3$ is of special interest. The Derdzinski metrics are then conformally flat.

The author proved ([1]) that, on a compact orientable 4-manifold with non vanishing signature, any metric with harmonic curvature has to be Einstein.

The proof of this fact relies heavily on the special nature of the tangent bundle among vector bundles over a manifold. Connections on the tangent bundle can be torsion free (as the Levi-Civita connection D is). In such a case their curvatures satisfy the first Bianchi identity. Since for the tangent bundle the same group acts on the form part of the curvature and on its Lie algebra valued part the decomposition into irreducible components of the curvature R^D is a refinement of the general decomposition of the curvature of an SO_4 connection on an oriented 4-manifold : one has

$$R^D = W^+ + W^- + Z + \frac{1}{6} u \, Id_{\Lambda^2 T}$$

with

$$R^+_- = {}^tR^-_+ = Z$$

$$R^+_+ = W^+ + \frac{1}{6} u \, Id_{\Lambda^+ T}$$

$$R^-_- = W^- + \frac{1}{6} u \, Id_{\Lambda^- T} .$$

Here W denotes the Weyl curvature tensor and $W^{\pm}$ its restriction to $\Lambda^{\pm} T$. The component Z is determined by the traceless part of the Ricci tensor.

Notice that the vanishing of R^+_- forces the vanishing of R^-_+ (and conversely) and is equivalent to the metric being Einstein.

The aforementioned theorem on harmonic curvature metrics on a manifold with non-vanishing signature can therefore be thought of as a twofold self-duality statement. Its main interest is that it has been obtained under a purely topological assumption.

Let us conclude by saying a few words on metrics of different signatures in particular Lorentzian metrics [for harmonic map theory, see [6] for instance]. Most of the geometric features of the theories discussed persist, but from the P.D.E point of view the systems become hyperbolic, hence show a completely different behavior. Moreover the whole context in which one looks for solutions is different : instead of starting with a given compact manifold M, one builds M while solving the equation. Very often one must deal with non compact spaces, having to specify what one means by asymptotically admissible data.

REFERENCES

[1] J.P. BOURGUIGNON, Les variétés riemanniennes de dimension 4 à signature non-nulle dont la courbure est harmonique sont d'Einstein, Inventiones Mat. 63 (1981) , 263-286.

[2] J.P. BOURGUIGNON, H.B. LAWSON, Stability and gap phenomena for Yang-Mills fields, Comm. in Mat. Phys. 79 (1981) , 189-230.

[3] J.P. BOURGUIGNON, H.B. LAWSON, Yang-Mills theory : its physical origins and differential geometric aspects, to appear in Ann. of Math. Studies, Princeton (1981).

[4] A. DERDZINSKI, Classification of certain compact Riemannian manifolds with harmonic curvature and non-parallel Ricci tensor, Math. Z. 172 (1980), 273-280.

[5] V.G. DRINFELD, Y.I. MANIN, A description of instantons, Comm. in Mat. Phys. 63 (1978) , 177-192.

[6] C.H. GU, On the harmonic maps from 2-dimensional space-time to Riemannian manifolds, Preprint ITP-SB (1980).

[7] J. ILIOPOULOS, Unified theories of elementary particle interactions, Contemporary Phys. 21 (1980), 159-183.

[8] C. LANCZOS , The splitting of the Riemann curvature tensor, Rev. Modern Phys. 34 (1962), 379-389.

[9] Y.L. XIN, Some results on stable harmonic maps, Preprint ITP SB (1980).

Harmonic Maps from $\mathbb{C}P^1$ to $\mathbb{C}P^n$*

D. Burns**

We present here a description of the classification of harmonic maps from $\mathbb{C}P^1$ to $\mathbb{C}P^n$ (with standard metrics). We learned at the Tulane conference that several other authors ([3], [5] and [6]) had worked out the classification about the same time, all by essentially the same methods, due originally to Calabi and Chern. Eells and Wood [5] have obtained interesting results about the harmon maps of higher genus surfaces to $\mathbb{C}P^n$ as well. Since the details of the (more general) proofs are so well-presented in [5], in this report we will content ourselves with the description of the classification and remarks on our point of view which are hopefully comple-mentary to Eells's lectures [3] at the conference, and [5]. We will start with the physical and geometric motivation for the classificat and end with some open questions which appear to us to be raised.

It's a pleasure to extend our thanks to R. Knill for his effort in organizing the Tulane conference, and to Jim Eells for several helpful comments on our manuscript.

*Lecture given at NSF-CBMS Regional Conference on Harmonic Maps, Tulane University, December 1980.

**A. P. Sloan Foundation Fellow. Research supported in part by NSF MCS-7900285.

1. Physical Motivation

Harmonic maps from $\mathbb{R}^2$ to S^2, and by 2-dimensional conformal invariance, from S^2 to S^2, are used to represent stationary energy states for an isotropic plane rotor system (usually referred to as the Heisenberg isotropic ferromagnet, or O(3) non-linear σ-model) and, with more general target space than S^2, has recently attracted a great deal of attention as the 2-dimensional analogue of classical Yang-Mills theories in dimension four. The main reasons for this appear to be that both theories are conformally invariant in one distinguished dimension d, and, closely related to this invariance by simple dimensional scaling considerations, admit solutions of finite energy on $\mathbb{R}^d$ in those distinguished dimensions only. According to now standard physics procedures, one would like to know the classical solutions, harmonic maps or Yang-Mills connections, as the starting point of an approximation to the full (Euclidean) quantum theory. For a cogent overview of this, c.f. chap. I of [7]. We considered harmonic maps from $S^2 = \mathbb{C}P^1$ to $\mathbb{C}P^n$ to try to understand the following question about Yang-Mills solutions on S^4.

Recall the (mathematical) set-up for the Yang-Mills equations. Let E be a complex vector bundle on S^4, with fixed hermitian matric (,). Consider all connections ∇ on E compatible with the metric, and let $F = F(\nabla)$ be the curvature tensor of ∇. The (second-order) Yang-Mills equations are the Euler-Lagrange equations of the variational problem

$$\delta(\int |F|^2 \, d \text{ vol}) = 0 ,$$

the variation taken over all smooth ∇ , or perhaps more restricted ∇ if one considers the structure group of E further reduced. If ∇ satisfies $*F = \pm F$, ∇ is called $\pm$ self-dual, respectively. (Here $*$ is the Hodge star-operator on S^4 , acting on the two-form components of F.) The second-order Yang-Mills equations are trivially satisfied for ∇ , if ∇ is $\pm$ self-dual. These equations are first order in the components of ∇ . It is well-known now that under the Ward-Penrose correspondence, these first order equations on ∇ are equivalent to the integrability equations for introducing a holomorphic (or anti-holomorphic) structure on the pull-back of E to $\mathbb{C}P^3$ under the quaterniomic fibration $\mathbb{C}P^2 \to S^4 = \mathbb{H}P^1$.

If (E_+,∇_+), (E_-,∇_-) are $\pm$ self-dual solutions (as indicated), then $E = E_+ \oplus E_-$, $\nabla = \nabla_+ \oplus \nabla_-$ is a solution of the Yang-Mills equations which is not $\pm$ self-dual. It is an open question as to whe any solution E , ∇ of the second-order equations <u>on</u> <u>S^4</u> is express as such a sum. (This is <u>false</u> if S^4 is replaced by, e.g., $S^1 \times S^3$)

2. Some Analogies

We consider instead the analogous but easier problem of trying to decompose a harmonic map $f : S^2 \to \mathbb{C}P^n$ into a "sum" of two harmonic maps f_+ and f_- , where $f_\pm$ is $\pm$ holomorphic (-holomorph being anti-holomorphic) of $\mathbb{C}P^1$ into Kahler target manifolds $M_\pm$. Because of the gauge-invariance of the problem under the action of $U(n+1)$ on $\mathbb{C}P^n$, $M_\pm$ should also be homogeneous spaces for $U(n+1)$, i.e., complex grassmannians. In order to guess what the "sum" of a holomorphic map and an anti-holomorphic map should be, let us pursue the analogy with Yang-Mills and vector bundles. We consider a harmonic map as giving a harmonic cycle in the target manifold, together with a preferred parmetrization. In algebraic geometry, there is a well-known correspondence between vector bundles and cycles, assigning to a vector bundle the cycle of zeros of a generic section. The direct-sum of vector bundles, as in the Yang-Mills question posed above, should therefore correspond to the intersection of cycles (the common zeroes of two sections). Thus, for our "sum" operation for our sought maps f_+ and f_- we should look for a sort of intersection operation. We now make these analogies precise.

3. Gauss-maps and intersection pairing

If $f : \mathbb{C}P^1 \to \mathbb{C}P^n$ were holomorphic, then there would be a canonical $U(n+1)$-equivariant way to associate to f mappings into $Gr(n+1, p+1)$, the grassmannian of complex $p+1$-planes in $\mathbb{C}^{n+1}$, namely, the osculating curves of f. The pth-osculating curve $f^{(p)} : \mathbb{C}P^1 \to Gr(n+1, p+1)$ associates to $z \in \mathbb{C}P^1$, the projective p-plane in $\mathbb{C}P^n$ through $f(z)$ which osculates $f(\mathbb{C}P^1)$ to order p at $f(z)$. In homogeneous coordinates, if $f = (f_o(z),\ldots,f_n(z))$, then $f^{(p)}(z)$ is the homogeneous $(p+1)$-plane spanned by $(\frac{d^k}{dz^k} f_o,\ldots,\frac{d^k}{dz^k} f_n)$, $k = 0,1,\ldots,p$. The maps $f^{(p)}$ are clearly holomorphic. One has similar constructions for anti-holomorphic curves.

We are interested in the case where f is harmonic, but not $\pm$ holomorphic. Then we can consider osculating maps $f_{\pm}^{(p)}$, $f_+^{(p)}(z)$ spanned by

$$\left(\frac{\partial^k f_o}{\partial z^k}, \ldots, \frac{\partial^k f_n}{\partial z^k}\right), \qquad k=0,1,\ldots,p$$

and $f_-^{(p)}(z)$ spanned by

$$\left(\frac{\partial^k f_o}{\partial \bar{z}^k}, \ldots, \frac{\partial^k f_n}{\partial \bar{z}^k}\right), \qquad k=0,1,\ldots,p .$$

We can now state how to decompose a harmonic map $f : \mathbb{C}P^1 \to \mathbb{C}P^n$ into two $\pm$ holomorphic maps.

<u>Theorem</u>. Let $f : \mathbb{C}P^1 \to \mathbb{C}P^n$ be a harmonic map whose image is not contained in any hyperplane. Then there exist integers $p,q \geq 0$ with $p+q = n$ such that

(1) $f_+^{(p)} : \mathbb{C}P^1 \to Gr(n+1, p+1)$ is anti-holomorphic

(2) $f_-^{(q)} : \mathbb{C}P^1 \to Gr(n+1, q+1)$ is holomorphic.

Furthermore, $f(z) = f_+^{(p)}(z) \cap f_-^{(q)}(z) \in \mathbb{C}P^n$, and $f_+^{(p)}(z)$, $f_-^{(q)}(z)$ intersect orthogonally at $f(z)$.

We call (p,q) the type of f. As an example, if f is already holomorphic, p will be n, $q = 0$, $f_+^{(n)}$ is a constant map, and $f_-^{(0)} = f$.

The proof, as already mentioned, uses the technique of Calabi [1] to show that $f_+^{(k)}(z)$ intersects $f_-^{(\ell)}(z)$ <u>orthogonally</u> at at $f(z)$ for any k,ℓ . One continues taking z or $\bar{z}$ derivatives of f until $k+\ell = n$. This maximal pair (k,ℓ) is (p,q), and, for example, since $\left(\frac{\partial^{p+1}}{\partial z^{p+1}} f_o, \ldots, \frac{\partial^{p+1}}{\partial z^{p+1}} f_n\right)$ must be in $f_+^{(p)}(z)$

by a simple dimension count and orthogonality, $f_+^{(p)}$ is anti-holomorphic.

We note that conversely, if g_+, g_- are $\mp$ holomorphic from $\mathbb{C}P^1$ to $Gr(n+1, p+1)$, $Gr(n+1, q+1)$, respectively, then $g(z) = g_+(z) \cap g_-(z) \in \mathbb{C}P^n$ defines a harmonic map if and only if $g_+(z)$ and $g_-(z)$ intersect orthogonally at $g(z)$. This necessary coupling conditions on g_+ and g_- to yield a non $\pm$ holomorphic harmonic map is in marked contrast to the situation in §1 for Yang-Mills fields, where arbitrary solutions $E_{\pm}$, $\nabla_{\pm}$ of the $\pm$ self-dual equations could be added to form a non $\pm$ self-dual solution. This added rigidity for harmonic maps may be because the gauge group is finite dimensional. Using the orthogonality relations and classical facts about curves in grassmannians, one finds:

Theorem (bis): f, $f_+^{(p)}$, $f_-^{(q)}$ as above. Then there exists a holomorphic map $g : \mathbb{C}P^1 \to \mathbb{C}P^n$ such that $g^{(q)} = f_-^{(q)}$. Furthermore, let $h : \mathbb{C}P^1 \to \mathbb{C}P^n$ be the composition of $g^{(n-1)} : \mathbb{C}P^1 \to Gr(n+1,n)$ with the anti-holomorphic map $Gr(n+1,n) \to Gr(n+1,1) = \mathbb{C}P^n$ sending a hyperplane to its orthocomplement. h is anti-holomorphic, and $h^{(p)} = f_+^{(p)}$.

Thus, $f_+^{(p)}$ is completely determined by $f_-^{(q)}$ and vice-versa!

4. Concluding Remarks.

If we call the holomorphic map g of Theorem (bis) the generating map for f, we see that each holomorphic map g generates $n+1$ harmonic maps, of all types (p,q), $p+q = n$. More specifically, given a holomorphic $g = (g_o(z),\dots,g_n(z))$, compute $h = (h_o(z),\dots,$

$h_n(z))$ satisfying

$$\sum_{i=0}^{n} \left(\frac{d^k}{dz^k} g_i\right) h_i = 0 \ , \ k = 0,1,\ldots,n-1 \ .$$

The harmonic map f of type (p,q) generated by g has local representation:

$$n\times n \text{ - minors of } \begin{bmatrix} g_o & \cdots & g_n \\ \frac{dg_o}{dz} & \cdots & \frac{dg_n}{dz} \\ \cdots & \cdots & \cdots \\ \frac{dg_o}{dz^{p-1}} & \cdots & \frac{dg_n}{dz^{p-1}} \\ \bar{h}_o & \cdots & \bar{h}_n \\ \left(\frac{\overline{dh_o}}{dz}\right) & \cdots & \left(\frac{\overline{dh_n}}{dz}\right) \\ \cdots & \cdots & \cdots \\ \left(\frac{\overline{dh_o}}{dz^{q-1}}\right) & \cdots & \left(\frac{\overline{dh_n}}{dz}\right) \end{bmatrix} .$$

We make two remarks on this formula. First, we have given a kind of Bäcklund transformation constructing non-holomorphic solutions of the harmonic map equation from holomorphic ones. Second, the type (p,q) of f seems very analogous to the Hodge (p,q)-type of harmonic forms on a Kahler manifold. Such a harmonic form of type (p,q) is locally a sum of products of p one-forms of holomorphic type, and q of anti-holomorphic type. In our case, the components are literally given as a sum of products of p holomorphic functions and q anti-holomorphic functions. Is there any general reason for these two phenomena? Of course, our example is very special.

Following Calabi, one can calcualte the energy of f in terms of the degrees of $f_+^{(p)}$, $f_-^{(q)}$, or in terms of the generating function g, by the classical Plücker formulas. Eells and Wood have estimated the index of the non-holomorphic harmonic $f: \mathbb{C}P^1 \to \mathbb{C}P^n$ in [5], again using complex curve theory. The reader should refer to [5] for details, as well as further references to the recent literature on the problem.

In conclusion, I should say that the formalism here was dictated by the analogy to the Yang-Mills problem. It is not clear that it can be generalized significantly for harmonic maps, e.g., for maps of S^2 to more general targets. The variant of Chern [2] may require less on the curvature of a target manifold to yield interesting results.

Mathematics Department
University of Michigan
Ann Arbor, Michigan 48109
U.S.A.

REFERENCES

[1] E. Calabi, Minimal immersions of surfaces in Euclidean spheres, J. Diff. Geom. 2 (1967), 111-125.

[2] S. S. Chern, On the minimal immersions of two-sphere in a space of constant curvature. Problems in Analysis, Princeton Univ. Press (1960), 27-40.

[3] A. M. Din, W. J. Zakrzewski, General Classical Solutions in the $\mathbb{C}P^{n-1}$ model, Nucl. Phys. B 174 (1980), 397-406.

[4] J. Eells, L. Lemaire, Selected topics in harmonic maps, this volume.

[5] J. Eells, J. C. Wood, Harmonic maps from surfaces to complex projective spaces, preprint.

[6] V. Glaser, R. Stora, Regular solutions of the $\mathbb{C}P^n$ models and further generalizations, preprint.

[7] A. Jaffe, C. Taubes, Vortices and Monopoles, Birkhäuser (1980).

VECTOR CROSS PRODUCTS, HARMONIC MAPS AND THE CAUCHY RIEMANN EQUATIONS

Alfred Gray

1. INTRODUCTION. A well-known property of Kähler manifolds is that Kähler submanifolds are minimal varieties [W1]. This is a special case of the fact that holomorphic maps between Kähler manifolds are harmonic [LZ1,2]. The reason is that holomorphic maps satisfy the Cauchy-Riemann equations.

I shall generalize these results in the present note to manifolds with parallel vector cross products. The main interest in doing so is to clarify and generalize some results of Harvey and Lawson [HL1,2]. They showed that there are many interesting 3-dimensional and 4-dimensional minimal varieties of $\mathbb{R}^8$. I shall define the notion of vector-cross-product-preserving map, holomorphic map being a special case. Many of the results of [HL1,2] have a simple elegant formulation because the immersions considered by Harvey and Lawson are vector-cross-product-preserving.

The main results proved are the following.

Theorem 1.1. Let M^k and $\bar{M}^\ell$ be Riemannian manifolds each with an r-fold vector cross product. Let $\Phi : M^k \to \bar{M}^\ell$ be a C^∞ map that preserves the vector cross products.

(i) If $r = 1$, then Φ is a holomorphic map between almost complex manifolds.

(ii) If $r \geq 2$, then each tangent map Φ_{*m} is either zero or injective.

In (iii)-(v) assume that each tangent map Φ_{*m} is injective.

(iii) If $r = 2$, then Φ is an isometric immersion $\Phi : M^3 \to \bar{M}^7$ or a local isometry.

(iv) If $r = 3$, then Φ is an isometric immersion $\Phi : M^4 \to \bar{M}^8$ or a local isometry.

(v) If $r \geq 4$, then $\Phi : M^{r+1} \to \bar{M}^{r+1}$ is a local isometry.

Theorem 1.2. Assume the hypotheses of theorem 1.1 and in addition assume that the vector cross products of M^k and $\bar{M}^\ell$ are parallel. Then Φ is harmonic. In particular if Φ is an isometric immersion, and it is equivalent to that of [EL]

In the special case that Φ is an isometric immersion (that is when Φ_* is injective on each tangent space) the map $(X,Y) \to T_X Y$ is the second fundamental form of the immersion. In this case $\tau(\Phi)$ becomes the mean curvature vector H.

<u>Definition</u>. The map Φ is said to be <u>harmonic</u> if and only if $\tau(\Phi)$ vanishes.

This is the usual definition of harmonic as given for example in [EL]. When Φ is an isometric immersion, Φ is harmonic if and only if M is a minimal variety.

In this section and the next sufficient conditions for a vector-cross-product-preserving map to be harmonic are given. For references to harmonic maps see the recent survey article [EL].

Before proving theorem 1.2 some algebraic facts about vector cross products will be needed.

<u>Lemma 4.1</u>. Let P be an r-fold vector cross product. Then

$$(4.2)\quad P(x_1\wedge\dots\wedge \overset{i}{P(x_1\wedge\dots\wedge x_r)}\wedge\dots\wedge x_r) = \sum_{h=1}^{r} (-1)^{i+h+1} \langle x_1\wedge\dots\wedge\hat{x}_i\wedge\dots\wedge x_r, x_1\wedge\dots\wedge\hat{x}_h\wedge\dots\wedge x_r\rangle x_h$$

$$(4.3)\quad P(x_1\wedge\dots\wedge \overset{i}{P(x_1\wedge\dots\wedge \overset{j}{y}\wedge\dots\wedge x_r)}\wedge\dots\wedge \overset{j}{z}\wedge\dots\wedge x_r) + P(x_1\wedge\dots\wedge \overset{i}{P(x_1\wedge\dots\wedge \overset{j}{z}\wedge\dots\wedge x_r)}\wedge\dots\wedge \overset{j}{y}\wedge\dots\wedge x_r)$$

$$\begin{aligned}
= \; & -2\langle x_1\wedge\dots\wedge\hat{x}_i\wedge\dots\wedge y\wedge\dots\wedge x_r, x_1\wedge\dots\wedge\hat{x}_i\wedge\dots\wedge z\wedge\dots\wedge x_r\rangle x_i \\
& + (-1)^{i+j+1}\{\langle x_1\wedge\dots\wedge\hat{x}_i\wedge\dots\wedge y\wedge\dots\wedge x_r, x_1\wedge\dots\wedge\hat{x}_j\wedge\dots\wedge x_r\rangle z \\
& + \langle x_1\wedge\dots\wedge\hat{x}_i\wedge\dots\wedge z\wedge\dots\wedge x_r, x_1\wedge\dots\wedge\hat{x}_j\wedge\dots\wedge x_r\rangle y\} \\
& + \sum_{h\neq i,j} (-1)^{h+i+1}\{\langle x_1\wedge\dots\wedge\hat{x}_i\wedge\dots\wedge \overset{j}{y}\wedge\dots\wedge x_r, x_1\wedge\dots\wedge \overset{j}{z}\wedge\dots\wedge\hat{x}_h\wedge\dots\wedge x_r\rangle \\
& + \langle x_1\wedge\dots\wedge\hat{x}_i\wedge\dots\wedge \overset{j}{z}\wedge\dots\wedge x_r, x_1\wedge\dots\wedge y\wedge\dots\wedge\hat{x}_h\wedge\dots\wedge x_r\rangle\} x_h .
\end{aligned}$$

<u>Proof</u>. To prove (4.2) one uses a polarized version of (2.2) together with the rule for expanding determinants. The result is

then Φ is a minimal variety.

Vector cross products were invented by Eckmann [E1,2] and can be considered to be generalizations of both the original Gibbs vector cross product and also of almost complex structures:

Definition. Let V be a finite dimensional vector space of dimension n over a field $\mathbb{F}$, and let $<,>$ be a symmetric bilinear form on V. An r-fold vector cross product on V is a multilinear map $P : \underbrace{V\times\ldots\times V} \to V$ such that

(i) $\|P(v_1,\ldots,v_r)\|^2 = \det(<v_i,v_j>) = \|v_1\wedge\ldots\wedge v_r\|^2$ (where $\|v\|^2 = <v,v>$) and

(ii) $<P(v_1,\ldots,v_r),v_i> = 0$ for $i = 1,\ldots,r$ and $v_1,\ldots,v_r \in V$.

In the present note I shall always assume that $\mathbb{F} = \mathbb{R}$ and that $<,>$ is positive definite. For the other cases see [BG] and [GR1]. Certain results in this paper are true when $<,>$ is indefinite.

Eckmann [E1,2] observed that it is possible to classify completely vector cross products. There are four types:

Type I: the $(n-1)$-fold product on $\mathbb{R}^n$;

Type II: the 1-fold product on $\mathbb{R}^{2n}$;

Type III: the 2-fold products on $\mathbb{R}^3$ and $\mathbb{R}^7$;

Type IV: the 3-fold product on $\mathbb{R}^4$ and the two 3-fold products on $\mathbb{R}^8$.

It should be pointed out that Eckmann [E1,2] originally required continuity in his definition of vector cross product. The classification is the same [E1,2], [WH].

It makes sense to consider a differentiable manifold having a vector cross product on each tangent space such that the vector cross product varies in a differentiable way [GR1,3]. Such a manifold is a natural generalization of an almost complex manifold. The generalization of holomorphic map is a vector-cross-product-preserving map. For $r > 1$ there are generalized Cauchy-Riemann equations for the vector cross product, but they are nonlinear.

As a manifold $\mathbb{R}^7$ has a 2-fold parallel vector cross product. Similarly $\mathbb{R}^8$ has two 3-fold parallel vector cross products. Consequently theorems 1.1 and 1.2 apply to vector-cross-product preserving maps into these spaces. These are the situations treated in [HL1,2].

In section 4 manifolds will be considered whose vector cross products are not parallel, but still have nice properties. For example, Lichnerowicz [LZ1,2] has shown that holomorphic maps between certain almost Hermitian manifolds are harmonic. Lichnerowicz's result implie that a holomorphic map from a Riemann surface into S^6 is harmonic, for example. In section 4 I shall give theorems that include this fa and also imply that a vector-cross-product-preserving map from an orientable 3-dimensional manifold in S^7 is harmonic, and in fact is a minimal variety.

2. VECTOR CROSS PRODUCTS ON RIEMANNIAN MANIFOLDS. Let M^n be an n-dimensional C^∞-Riemannian manifold with metric tensor $<,>$. Denote by $\mathfrak{X}(M)$ the Lie algebra of C^∞ vector fields on M. Tensor fields of type (r,1) will be regarded as maps $\mathfrak{X}(M)\times\ldots\times\mathfrak{X}(M) \to \mathfrak{X}(M)$ that are linear with respect to functions. Then an r-fold vector cross product on M can be regarded as a tensor field P of type (r,1) such that

$$(2.1) \quad <P(X_1,\ldots,X_r),X_i> = 0 \quad \text{for} \quad i = 1,\ldots,r;$$

$$(2.2) \quad \|P(X_1,\ldots,X_r)\|^2 = \|X_1\wedge\ldots\wedge X_r\|^2 \quad (=\det(<X_i,X_j>)),$$

for $X_1,\ldots,X_r \in \mathfrak{X}(M)$.

To each r-fold vector cross product there is an associated (r+1)-form φ given by

$$(2.3) \quad \varphi(X_1,\ldots,X_{r+1}) = <P(X_1,\ldots,X_r),X_{r+1}>.$$

φ is called the fundamental form of P. Property (2.1) implies that φ and P are antisymmetric. To emphasize this fact I shall write $P(X_1\wedge\ldots\wedge X_r)$ for $P(X_1,\ldots,X_r)$ and $\varphi(X_1\wedge\ldots\wedge X_{r+1})$ for $\varphi(X_1,\ldots,X_{r+1})$.

There are two kinds of obstructions to the existence of vector cross products on manifolds. One comes from the linear algebra of a vector cross product at a point; there are also global topological obstructions to the existence of a vector cross product [GR1,3], [GG].

The descriptions of the four types of vector cross products on manifolds are as follows:

Type I: $r = n - 1$. This kind of vector cross product coincides with the Hodge operator $*$ operating on (n-1)-vectors.

Type II: $r = 1$, n even. In this case the vector cross product is the same as an almost complex structure compatible with the metric. The fundamental form is just the Kähler form. Usually J is written instead of P, and F instead of φ in this case.

Type III: $r = 2$, $n = 3$ or 7. When $r = 2$, $n = 3$, a type III vector cross product is the ordinary Gibbs vector cross product (and it is also of type I). For $r = 2$, $n = 7$, a type III vector cross product is defined using the Cayley numbers in exactly the same way that the Gibbs product is defined using the quaternions [C], [BG], [E1,2], [GR1,3].

Type IV: r = 3, n = 4 or 8. When r = 3, n = 4, a type IV vector cross product is already of type I. The case r = 3, n = 8 the most esoteric of the vector cross products. In fact $\mathbb{R}^8$ has tw nonisomorphic vector cross products of type IV. See [BG], [GR1,3], [for functions using the Cayley numbers. There are two nonisomorphic products because of the nonassociativity of the Cayley numbers. The two vector cross products are related by the triality automorphism Spin(8) [BG], [GR1,3], [GG].

Suppose that M has a vector cross product on each tangent spac There are still global obstructions to the existence of a continuous vector cross product on all of M. For example M must be orientabl This is sufficient for the existence of a type I vector cross product but for the other types further obstructions exist. Most of these are expressible in terms of the Stiefel Whitney classes of M. See [GR1], [GG].

I turn now to the differential geometry of vector cross products The most natural condition to impose on a vector cross product on a Riemannian manifold is that it be parallel. (However weaker conditio will be considered in section 4.) Type I vector cross products are automatically parallel. To say that a type II product (that is an almost complex structure) is parallel is the same thing as saying tha M is a Kähler manifold. The natural 2- and 3-fold vector cross prod on $\mathbb{R}^7$ and $\mathbb{R}^8$ (considered as manifolds) are parallel. In fact t existence of a parallel type III vector cross product on M^7 or a pa lel type IV vector cross product on M^8 is a good geometric way of s ing that the holonomy group of M is a subgroup of G_2 or Spin(7) [GR1,3]. Moreover, all such manifolds are Ricci flat [BO]. The auth is unaware of any examples of nonflat manifolds (even locally) with parallel vector cross products of type III or IV.

3. INJECTIVITY OF VECTOR-CROSS-PRODUCT-PRESERVING MAPS. This section concerns linear algebra only, except at the end where theorem 1.1 is proved. Let V^k and $\bar{V}^\ell$ be vector spaces over $\mathbb{R}$, each with a positive definite inner product which will be denoted by $<,>$, and each with a vector cross product which will be denoted by P.

Definition. A linear transformation $A : V^k \to \bar{V}^\ell$ is said to be vector-cross-product-preserving provided

$$AP(x_1 \wedge \ldots \wedge x_r) = P(Ax_1 \wedge \ldots \wedge Ax_r)$$

for all $x_1, \ldots, x_r \in V$.

Lemma 3.1. Assume $r \geq 2$ and let A be a vector-cross-product-preserving linear transformation between vector spaces V^k and $\bar{V}^\ell$, each having an r-fold vector cross product. Then A is either zero or injective.

Proof. Suppose A is not injective. Let $x \in V^k$ with $x \neq 0$ but $Ax = 0$. Then $AP(x \wedge x_2 \wedge \ldots \wedge x_r) = 0$ for all $x_2, \ldots, x_r \in V^k$. But any vector perpendicular to x can be written in the form $P(x \wedge x_2 \wedge \ldots \wedge x_r)$. It follows that $Ay = 0$ for all $y \in V^k$.

Lemma 3.2. Assume $r \geq 2$ and let A be a vector-cross-product-preserving linear transformation between vector spaces V^k and $\bar{V}^\ell$ with r-fold vector cross products. Assume $A \neq 0$. Then A preserves the lengths of decomposable $(r-1)$-vectors.

Proof. Let $z, x_1, \ldots, x_r \in V^k$. Then

$$\begin{aligned} (3.1)\quad & <P(x_1 \wedge \ldots \wedge x_{r-1} \wedge z), P(x_1 \wedge \ldots \wedge x_{r-1} \wedge P(x_1 \wedge \ldots \wedge x_r))> \\ &= <x_1 \wedge \ldots \wedge x_{r-1} \wedge z, x_1 \wedge \ldots \wedge x_{r-1} \wedge P(x_1 \wedge \ldots \wedge x_r)> \\ &= \|x_1 \wedge \ldots \wedge x_{r-1}\|^2 <z, P(x_1 \wedge \ldots \wedge x_r)>. \end{aligned}$$

Since (3.1) holds for arbitrary z it follows that

$$\begin{aligned} (3.2)\quad & P(x_1 \wedge \ldots \wedge x_{r-1} \wedge P(x_1 \wedge \ldots \wedge x_{r-1} \wedge P(x_1 \wedge \ldots \wedge x_r))) \\ &= -\|x_1 \wedge \ldots \wedge x_{r-1}\|^2 P(x_1 \wedge \ldots \wedge x_r) \end{aligned}$$

for all $x_1, \ldots, x_r \in V^k$. By lemma 3.1, A is injective. Hence from

(3.2) it follows that

$$\|A(x_1\wedge\ldots\wedge x_r)\|^2 = \|x_1\wedge\ldots\wedge x_r\|^2.$$

Lemma 3.3. Let V^k and $\bar{V}^\ell$ be finite dimensional vector spaces over $\mathbb{R}$, each with a positive definite inner product $<,>$. Let $0 < p < k$. Assume that $A : V^k \to \bar{V}^\ell$ is a linear transformation such that

$$(3.3)\qquad \|A(x_1\wedge\ldots\wedge x_p)\|^2 = \|x_1\wedge\ldots\wedge x_p\|^2$$

for all $x_1,\ldots,x_p \in V^k$. Then

$$(3.4)\qquad \|Ax\|^2 = \|x\|^2$$

for all $x \in V$.

Proof. Let $L = {}^tAA$ where ${}^tA$ denotes the adjoint of A. Then $L : V^k \to V^k$ is symmetric with respect to $<,>$. Hence it is diagonalizable and has eigenvalues $\lambda_1,\ldots,\lambda_k$. Now (3.3) implies that

$$(3.5)\qquad <L(x_1\wedge\ldots\wedge x_p),x_1\wedge\ldots\wedge x_p> = \|x_1\wedge\ldots\wedge x_p\|^2$$

for all $x_1,\ldots,x_p \in V^k$. From (3.5) it is easy to see that

$$(3.6)\qquad \lambda_{i_1}\lambda_{i_2}\ldots\lambda_{i_p} = 1$$

for all $i_1,\ldots,i_p$, with $1 \le i_1 < i_2 <\ldots< i_p \le k$. Solving the equation given by (3.6) one finds that $\lambda_i = 1$ for all i. Hence (3.4) follows.

Let M_m denote the tangent space to a differentiable manifold M at $m \in M$.

Definition. Let M^k and $\bar{M}^\ell$ be Riemannian manifolds each with an r-fold vector cross product P. A map $\Phi : M^k \to \bar{M}^\ell$ is said to be vector-cross-product-preserving provided that each tangent map $\Phi_{*m} : M^k_m \to \bar{M}^\ell_{\Phi(m)}$ preserves the vector cross products.

Proof of theorem 1.1. By definition when $r = 1$, Φ is holomorphic. The rest follows from lemmas 3.1 and 3.3.

4. HARMONICITY OF MAPS BETWEEN MANIFOLDS WITH PARALLEL VECTOR CROSS PRODUCTS.

Definition. Let M^k and $\bar{M}^\ell$ be Riemannian manifolds and let $\Phi : M^k \to \bar{M}^\ell$ be a C^∞ map. Denote by ∇ and $\bar{\nabla}$ the Riemannian connections of M and $\bar{M}$ respectively. The second fundamental form T of Φ is given by

$$(T_X Y)_m = (\bar{\nabla}_{\Phi_* X} \Phi_* Y)_m - \Phi_{*m}((\nabla_X Y)_m)$$

where $X, Y \in \mathfrak{X}(M)$ are vector fields projectable onto vector fields $\Phi_*(X),\ \Phi_*(Y) \in \mathfrak{X}(\bar{M})$.

Thus T has values in $\mathfrak{X}(\bar{M})$. Note that if $X, Y \in \mathfrak{X}(M)$ are projectable so is $\nabla_X Y$. It is not difficult to see that T is tensorial in the sense that

$$T_{(f\circ\Phi)X}(g\circ\Phi)Y = fgT_X Y$$

for $X, Y \in \mathfrak{X}(M)$ and C^∞ functions $f, g : M \to \mathbb{R}$. Moreover because ∇ and $\bar{\nabla}$ both have torsion zero it is easy to see that

$$T_X Y = T_Y X$$

for $X, Y \in \mathfrak{X}(M)$.

Consequently the second fundamental form T gives rise at each point $m \in M$ to a symmetric map $M_m \times M_m \to M^*_{\Phi(m)}$ which will be written as $T_x y$ for $x, y \in M_m$. Here

$$(4.1) \qquad T_x y = (T_X Y)_m$$

where X, Y are projectable vector fields such that $X_m = x$ and $Y_m = y$. Because T is tensorial the definition (4.1) is independent of the choice of X and Y. Moreover (4.1) can be used to define $T_X Y$ whether or not X and Y are projectable.

Definition. The tension field $\tau(\Phi)$ of Φ is given by

$$\tau(\Phi) = \sum_{i=1}^{k} T_{E_i} E_i$$

where $\{E_1, \ldots, E_k\}$ is a local orthonormal frame field on M. This definition is independent of the choice of local orthonormal frame field,

$$(4.4) \quad -<P(x_1\wedge\ldots\wedge\overset{i}{P}(x_1\wedge\ldots\wedge x_r)\wedge\ldots\wedge x_r,z>$$

$$= \quad <P(x_1\wedge\ldots\wedge\overset{i}{z}\wedge\ldots\wedge x_r),P(x_1\wedge\ldots\wedge x_r)>$$

$$= \quad <x_1\wedge\ldots\wedge\overset{i}{z}\wedge\ldots\wedge x_r,x_1\wedge\ldots\wedge x_r>$$

$$= \quad \sum_{h=1}^{r}(-1)^{i+h}<x_1\wedge\ldots\wedge\hat{x}_i\wedge\ldots\wedge x_r,x_1\wedge\ldots\wedge\hat{x}_h\wedge\ldots\wedge x_r><x_h,z>.$$

Since (4.4) holds for arbitrary z, (4.2) follows. Then (4.3) is jus
a polarized version of (4.2).

For $r \leq 3$ equation (4.2) written out is as follows:

$r = 1$: $J^2 = -I$ (writing $P=J$);

$r = 2$: $P(x\wedge P(x\wedge y)) = -\|x\|^2 y + <x,y>x$;

$r = 3$: $P(x\wedge y\wedge P(x\wedge y\wedge z)) = -\|x\wedge y\|^2 z + <x\wedge y,x\wedge z>y - <x\wedge y,y\wedge z>x$.

Proof of theorem 1.2. Here it will be assumed that Φ is an iso
metric immersion. The only other case of interest, that of holomorph:
maps, can easily be taken care of separately. Anyway holomorphic map
will be dealt with in the next section in a more general context.

Because the vector cross products of M and $\bar{\text{M}}$ are assumed to
be parallel one has

$$(4.5) \quad T_X P(Y_1\wedge\ldots\wedge Y_r) = \sum_{a=1}^{r} P(Y_1\wedge\ldots\wedge T_X Y_a\wedge\ldots\wedge Y_r)$$

for $X,Y_1,\ldots,Y_r \in \mathfrak{X}(M)$. Let $\{E_1,\ldots,E_k\}$ be a local orthonormal fra
field on M. Then from (4.1) and (4.2) it follows that

$$(4.6) \quad T_{P(E_1\wedge\ldots\wedge E_r)}P(E_1\wedge\ldots\wedge E_r)$$

$$= \quad \sum_{a=1}^{r} P(E_1\wedge\ldots\wedge T_{P(E_1\wedge\ldots\wedge E_r)}E_a\wedge\ldots\wedge E_r)$$

$$= \quad \sum_{a,b=1}^{r} P(E_1\wedge\ldots\wedge P(E_1\wedge\ldots\wedge T_{E_a}E_b\wedge\ldots\wedge E_r)\wedge\ldots\wedge E_r)$$

$$= \quad A + B.$$

Here (using (4.2) and the fact that $<T_{E_a}E_b,E_c>=0$)

$$A = \sum_{a=1}^{r} P(E_1\wedge\ldots\wedge \overset{a}{P(E_1\wedge\ldots\wedge T_{E_a}E_a\wedge\ldots\wedge E_r)}\wedge\ldots\wedge E_r)$$

$$= \sum_{a=1}^{r}\sum_{\substack{b=1\\ b\neq a}}^{r} (-1)^{a+b+1}\langle E_1\wedge\ldots\wedge\hat{E}_a\wedge\ldots\wedge E_r, E_1\wedge\ldots\wedge T_{E_a}E_a\wedge\ldots\wedge\hat{E}_b\wedge\ldots\wedge E_r\rangle E_b$$

$$= -\sum_{a=1}^{r} \|E_1\wedge\ldots\wedge\hat{E}_a\wedge\ldots\wedge E_r\|^2 T_{E_a}E_a$$

$$= -\sum_{a=1}^{r} T_{E_a}E_a .$$

Also using (4.3) and the fact that $\langle T_{E_a}E_b, E_c\rangle = 0$ one finds that

$$B = \sum_{a<b}\{P(E_1\wedge\ldots\wedge P(E_1\wedge\ldots\; T_{E_a}E_b\wedge\ldots\wedge E_r)\wedge\ldots\wedge E_b\wedge\ldots\wedge E_r)$$

$$+ P(E_1\wedge\ldots\wedge E_a\wedge\ldots\wedge P(E_1\wedge\ldots\wedge T_{E_a}E_b\wedge\ldots\wedge E_r)\}$$

$$= -\sum_{a<b}\{P(E_1\wedge\ldots\wedge \overset{a}{P(E_1\wedge\ldots\wedge \overset{a}{T_{E_a}E_b}\wedge\ldots\wedge \overset{b}{E_a}\wedge\ldots\wedge E_r)}\wedge\ldots\wedge \overset{b}{E_b}\wedge\ldots\wedge E_r)$$

$$+ P(E_1\wedge\ldots\wedge \overset{a}{P(E_1\wedge\ldots\wedge \overset{a}{T_{E_a}E_b}\wedge\ldots\wedge \overset{b}{E_b}\wedge\ldots\wedge E_r)}\wedge\ldots\wedge \overset{b}{E_a}\wedge\ldots\wedge E_r)\}$$

$$= 0.$$

Thus (4.6) reduces to

$$(4.7)\qquad T_{P(E_1\wedge\ldots\wedge E_r)}P(E_1\wedge\ldots\wedge E_r) = -\sum_{a=1}^{r} T_{E_a}E_a .$$

Using (4.7) the mean curvature vector H can be computed as follows.

$$(4.8)\qquad rH = r\sum_{i=1}^{k} T_{E_i}E_i = \sum_{i=1}^{r}\sum_{a_i=1}^{k} T_{E_{a_i}}E_{a_i}$$

$$= \sum_{a_1,\ldots,a_r=1}^{k} T_{P(E_{a_1}\wedge\ldots\wedge E_{a_r})}P(E_{a_1}\wedge\ldots\wedge E_{a_r})$$

$$= -\sum_{a_1,\ldots,a_{r-1}=1}^{k}\{H - T_{E_{a_1}}E_{a_1} - \ldots - T_{E_{a_{r-1}}}E_{a_{r-1}}\}$$

$$= -k^{r-1}H + k^{r-2}(r-1)H.$$

Thus from (4.8) it follows that

$$(4.9) \qquad \{k^{r-2}(k-r+1)+r\}H = 0.$$

Since the coefficient of H in (4.9) is always positive it follows that $H = 0$.

5. VECTOR-CROSS-PRODUCT-PRESERVING MAPS BETWEEN MANIFOLDS WITH NONPARALLEL VECTOR CROSS PRODUCTS. Sometimes it is possible to conclude that vector-cross-product-preserving maps are harmonic even though the vector cross products are not parallel. Certain weaker conditions suffice. This is the subject of the present section.

Let P be an r-fold vector cross product on a Riemannian manifold M and let φ be the associated $(r+1)$-form. Then

$$(5.1)\quad \nabla_Y(\varphi)(X_1\wedge\ldots\wedge X_{r+1}) = \langle\nabla_Y(P)(X_1\wedge\ldots\wedge X_r),X_{r+1}\rangle$$

$$(5.2)\quad \langle\nabla_Y(P)(X_1\wedge\ldots\wedge X_r),P(X_1\wedge\ldots\wedge X_r)\rangle = 0$$

for all $Y, X_1,\ldots,X_{r+1} \in \mathfrak{X}(M)$. See [GR1].

Definition. Let $\Phi : M^k \to \bar{M}^\ell$ be vector-cross-product-preserving. For tangent vectors $x,y_1,\ldots,y_r$ at a point $m \in M^k$ let

$$T_x(P)(y_1\wedge\ldots\wedge y_r) = T_xP(y_1\wedge\ldots\wedge y_r) - \sum_{a=1}^{r} P(\Phi_*y_1\wedge\ldots\wedge T_xy_a\wedge\ldots\wedge\Phi_*y_r).$$

Lemma 5.1. The second fundamental form T of a vector-cross-product-preserving map Φ satisfies

$$\bar{\nabla}_{\Phi_*x}(P)(\Phi_*y_1\wedge\ldots\wedge\Phi_*y_r) = T_x(P)(y_1\wedge\ldots\wedge y_r) + \Phi_*\nabla_x(P)(y_1\wedge\ldots\wedge y_r)$$

for $x,y_1,\ldots,y_r \in M_m$ and $m \in M$.

Proof. This is immediate from the definitions.

Each of the four kinds of vector cross products will be considered. However type I vector cross products can be excluded immediately because they are always parallel. (The associated form φ is just the volume form in this case.)

As for an almost complex structure J there is the following result of Lichnerowicz [LZ1,2]. (However, the proof given below is different from that of Lichnerowicz.)

Let F be the Kähler form of J and $\langle,\rangle$ and let δF be the coderivative of F. The vector field dual to δF will be denoted by $(\delta F)^*$.

Lemma 5.2. Let M^{2k} and $\bar{M}^{2\ell}$ be almost Hermitian manifolds. Let J and $\bar{J}$ be the almost complex structures and F and $\bar{F}$ the corresponding Kähler forms. Suppose $\Phi : M \to \bar{M}$ is holomorphic. Then

(5.3) $2J\tau(\Phi)$

$$= 2\Phi_*((\delta F)^*) + \sum_{i=1}^{2k} \{\bar{\nabla}_{\Phi_* E_i}(\bar{J})(\Phi_* E_i) + \bar{\nabla}_{J\Phi_* E_i}(\bar{J})J\Phi_* E_i\}$$

where $\{E_1, \ldots, E_{2k}\}$ is any local frame on M.

Proof. Let $X, Y \in \mathfrak{X}(M)$. Then

(5.4) $$T_X(J)Y + T_{JX}(J)JY = \bar{\nabla}_{\Phi_* X}(J)\Phi_* Y + \bar{\nabla}_{\Phi_* JX}(J)\Phi_* JY - \Phi_* \nabla_X(J)Y - \Phi_* \nabla_{JX}(J)JY.$$

Letting $E_i = X = Y$ in (5.4) and summing it follows that

(5.5) $$\sum_{i=1}^{2k} \{\bar{\nabla}_{\Phi_* E_i}(\bar{J})(\Phi_* E_i) + \bar{\nabla}_{J\Phi_* E_i}(\bar{J})J\Phi_* E_i\}$$

$$= 2 \sum_{i=1}^{2k} \{aT_{E_i}(J)E_i + \Phi_* \nabla_{E_i}(J)E_i\}$$

$$= 2 \sum_{i=1}^{2k} \{T_{E_i} JE_i - JT_{E_i} E_i\} - 2\Phi_*((\delta F)^*).$$

Now $\Sigma\, T_{E_i} JE_i = +\Sigma\, T_{J^2 E_i} J^2 E_i = -\Sigma\, T_{JE_i} E_i = \Sigma\, T_{E_i} JE_i$. Hence (5.5) reduces to (5.3).

Lemma 5.2 has the following consequence.

Theorem 5.3. Let M and $\bar{M}$ be almost Hermitian manifolds whose Kähler forms satisfy

(5.6) $\delta F = 0$

(5.7) $\bar{\nabla}_X(\bar{F})(Y \wedge Z) + \bar{\nabla}_{JX}(\bar{F})(JY \wedge Z) = 0$

for all $X, Y, Z \in \mathfrak{X}(M)$. Let $\Phi : M \to \bar{M}$ be holomorphic. Then Φ is harmonic.

Remarks. (1) In the terminology of [CR1] manifolds which satisfy (5.6) are called semi-Kählerian, and manifolds which satisfy (5.7) are called quasi-Kählerian.

(2) Condition (5.7) is equivalent to saying that the (2,1) and (1,2) parts of dF vanish.

(3) Examples of semi-Kählerian manifolds: any orientable 6-dimensional submanifold of $\mathbb{R}^8$; any compact homogeneous almost Hermitian manifold with positive Euler characteristic.

(4) Examples of quasi-Kählerian manifolds: Riemannian 3-symmetric spaces. The sphere S^6 is thus quasi-Kählerian. However many Riemannian 3-symmetric spaces including S^6 satisfy an even stronger condition: $\nabla_X(J)X = 0$. Such manifolds are called nearly Kählerian.

(5) A holomorphic map of Hermitian manifolds (that is complex manifolds whose metrics are compatible with the integrable almost complex structures) need not be harmonic [EL,p.38].

It is possible to generalize the notion of nearly Kähler to the other vector cross products.

Definition. Let P be an r-fold vector cross product on a Riemannian manifold. Then P is said to be nearly parallel provided

$$\nabla_X(P)(X \wedge Y_2 \wedge \ldots \wedge Y_r) = 0$$

for all $X, Y_2, \ldots, Y_r \in \mathfrak{X}(M)$.

As just remarked, S^6 and some other 3-symmetric spaces have nearly parallel almost complex structures. See [GR1,2,3] for a discussion of these manifolds.

In [GR1] it is shown that a type IV nearly parallel vector cross product must in fact be parallel. Therefore the only remaining case of interest is the type III nearly parallel vector cross products on 7-dimensional Riemannian manifolds.

In [GR1] it is shown that any orientable hypersurface of $\mathbb{R}^8$ has a 2-fold vector cross product. In particular the sphere S^7 has a 2-fold vector cross product, and this vector cross product turns out to be nearly parallel. Moreover Marchiafava has proved the converse [MF]:

Theorem 5.4. Let M^7 have a nearly parallel 2-fold vector cross product. Then M has constant nonnegative sectional curvature.

Now consider a vector-cross-product-preserving map $\Phi : M^3 \to \bar{M}^7$, where M^3 and $\bar{M}^7$ have 2-fold vector cross products. The vector cro product on M^3 is automatically parallel. Also by theorem 1.1, Φ must be an isometric immersion.

The following result will now be proved.

Theorem 5.5. Suppose the 2-fold vector cross product on $\bar{M}^7$ is nearly parallel. Then a vector-cross-product-preserving map $\Phi : M^3 \to$ must be harmonic so that M^3 is a minimal variety of $\bar{M}^7$.

Proof. Since Φ is an isometric immersion we omit Φ_* from the calculations. Then as a special case of (4.3) one has

$$(5.8)\qquad P(X\wedge P(Y\wedge T_X Y)) + P(Y\wedge P(X\wedge T_X Y)) = -2\langle X,Y\rangle T_X Y + \langle X,T_X Y\rangle Y + \langle Y,T_X Y\rangle$$

for $X,Y \in \mathfrak{X}(M)$. But for an isometric immersion X and Y are perpendicular to $T_X Y$ so that (5.8) reduces to

$$(5.9)\qquad P(X\wedge P(Y\wedge T_X Y)) + P(Y\wedge P(X\wedge T_X Y)) = -2\langle X,Y\rangle T_X Y.$$

To compute the mean curvature H let $X = E_i$ and $Y = E_j$ and sum (where $\{E_1,E_2,E_3\}$) is a local orthonormal frame on M). One finds that

$$(5.10)\qquad H = -\sum_{i,j=1}^{3} P(E_i\wedge P(E_j\wedge T_{E_i} E_j)).$$

On the other hand whether or not P is nearly parallel

$$\sum_{i=1}^{3} T_{E_i} P(E_i\wedge E_j) = \sum_{i=1}^{3} T_{P(E_i\wedge E_j)} P(P(E_i\wedge E_j)\wedge E_j)$$

$$= \sum_{i=1}^{3} T_{P(E_i\wedge E_j)} E_i$$

so that

$$(5.11)\qquad \sum_{i=1}^{3} T_{E_i} P(E_i\wedge Y) = 0$$

for all $Y \in \mathfrak{X}(M)$. Then (5.10) and (5.11) imply

$$(5.12)\qquad \sum_{i,j=1}^{3} P(E_i\wedge T_{E_i}(P)(E_i\wedge E_j)) = H.$$

Now one uses the hypotheses that P is nearly parallel together with (5.12) to conclude that $H = 0$.

REFERENCES

[BO] E. Bonan, "Sur des variétés riemanniennes à groupe d'holonomie G_2 ou Spin(7)," C.R. Acad. Sci. Paris 262 (1966), 127-129.

[BG] R. Brown and A. Gray, "Vector cross products," Comment. Math. Helv. 42 (1967), 222-236.

[C] E. Calabi, "Construction and properties of some 6-dimensional almost complex manifolds," Trans. Amer. Math. Soc. 87 (1958), 407-438.

[dR] G. deRham, "On the area of complex manifolds," Notes, Institute for Advanced Study, Princeton (1957). Part of these notes is contained in a paper of the same title in Global Analysis, papers in honor of K.Kodaira, Princeton, 1969, 141-148.

[E1] B. Eckmann, "Systeme von Richtungsfeldern in Sphären und stetige Lösungen komplexer linearer Gleichungen," Comment. Math. Helv. 15 (1943), 1-26.

[E2] B. Eckmann, "Continuous solutions of linear equations - some exceptional dimensions in topology," Batelle Rencontres (1967), Lectures in Mathematics and Physics, W.A. Benjamin (1968), 516-526.

[EL] J. Eells and L. Lemaire, "A report on harmonic maps," Bull. London Math. Soc. 10 (1978), 1-68.

[FG] M. Fernández and A. Gray, "Riemannian manifolds with structure group G_2" Ann. Math. Pura Appl. (to appear).

[GR1] A. Gray, "Vector cross products on manifolds," Trans. Amer. Math. Soc. 141 (1969), 465-504. Correction 148 (1970), 625.

[GR2] A. Gray, "Riemannian manifolds with geometric symmetries of order 3," J. Differential Geometry 7 (1972), 343-369.

[GR3] A. Gray, "The structure of nearly Kähler manifolds," Math. Ann. 223 (1976), 233-248.

[GR4] A. Gray, "Vector cross products," Rend. Sem. Mat. Univ. Politecn. Torino 35 (1976-1977), 69-75.

[GG] A. Gray and P. Green, "Sphere transitive structures and the triality automorphism," Pacific J. Math. 34 (1970), 83-96.

[GH] A. Gray and L.M. Hervella, "The sixteen classes of almost Hermitian manifolds and their linear invariants," Ann. Mat. Pura Appl. (IV) 123 (1980), 35-58.

[HL1] R. Harvey and H.B. Lawson, Jr., "A constellation of minimal varieties defined over the group G_2," Partial Differential Equations, Proceedings of the Park City Conference (1977), 167-187, Lecture Notes in Pure and Applied Mathematics vol.48, Marcel Dekker, New York, 167-187.

[HL2] R. Harvey and H.B. Lawson, Jr., "Geometries associated to the group SU_n and varieties of minimal submanifolds arising from Cayley arithmetic," preprint IHES, 1978.

[MF] S. Marchiafava, "Characterization of Riemannian manifolds with weak holonomy group G_2," preprint Università di Roma, 1980.

[LZ1] A. Lichnerowicz, "Sur les applications harmoniques," C.R. Acad. Sci. Paris 267 (1968), A548-553.

[LZ2] A. Lichnerowicz, "Applications harmoniques et variétés kählériennes," Sym. Math. III, Bologna (1970), 341-402.

[WH] G. Whitehead, "Note on cross-sections in Stiefel manifolds," Comment. Math. Helv. 37 (1962), 239-240.

[WI] W. Wintinger, "Eine Determinantenidentität unh ihre Anwendung auf analytische Gebilde in Euklidischen und Hermitescher Massbestimmung," Monatsch. f. Math. u. Physik 44 (1936), 343-365.

[Z] P. Zvengrowski, "A 3-fold vector product in $\mathbb{R}^8$," Comment. Math. Helv. 40 (1966), 149-152.

Department of Mathematics
University of Maryland
College Park, Maryland 20742
USA

HARMONIC MAPS IN KÄHLER GEOMETRY AND DEFORMATION THEORY

by

M. Kalka

1. Introduction:

Many of the great advances in complex analysis have come about by suitably expanding the class of objects under study from holomorphic objects to a somewhat larger class. Thus one studies k-quasiconformal mappings rather than restricting oneself to conformal mappings. More recently complex analysts have realized the profit in studying the inhomogeneous Cauchy-Riemann equations and not merely the homogeneous ones. This is particularly true when proving existence theorems. By enlarging the objects under study one has available the techniques of modern analysis, where one can then make local constructions which are impossible within the holomorphic category.

It was Riemann who first used variational theory, namely the Dirichlet principle, in the study of complex analysis. He showed that the biholomorphic equivalence of a simply connected plane domain (not all of $\mathbb{C}$) to the unit disc could be proved if one could prove an existence theorem for a certain variational problem. In the case of one complex variable this technique has proven itself and potential theory has become a standard tool. One of the reasons for the great success of potential theory in one variable is the ability to go backwards; that is, once one constructs a harmonic function it is an easy matter to construct the conjugate harmonic function (at least locally) and thereby a holomorphic function.

In more than one variable, or when one considers harmonic mappings between manifolds which have curved metrics matters are more complicated. If N,M are Kähler manifolds and $f : N \to M$ is a holomorphic (or conjugate holomorphic) map, then f is harmonic with respect to the Kähler metric on N, M. It is not true, however, that all harmonic maps are holomorphic or conjugate holomorphic. If $N \subseteq \mathbb{R}^3$ is an isometrically imbedded surface, then its Gauss map $\gamma : N \to S^2$ is harmonic if and only if N is imbedded as a surface of constant mean curvature. On the other hand, γ is conjugate holomorphic if and only if N is minimally imbedded. Thus if one takes a surface with constant but non-zero mean curvature, its Gauss map provides an example of a harmonic non-holomorphic map. For instance N could be taken to be the surface obtained by revolving, about the x-axis, the curve, in the x,y plane, traced out by a focus of a non-circular ellipse as the ellipse rolls along the x-axis.

The problem of determining what conditions are sufficient to guarantee that a harmonic map of Kähler manifolds is holomorphic has received much attention of late. The first result of this type is due to Eells-Wood [6].

Let $f : N \to M$ be a harmonic map of compact Riemann orientable surfaces, and denote their Euler characteristics by $\chi(N)$ and $\chi(M)$ respectivel Under these conditions Eells and Wood prove the following result.

Theorem: If $\chi(N) + |\deg(f)\, \chi(M)| > 0$, then f is either holomorphic or conjugate holomorphic.

As a corollary one obtains that for any pair of metrics on the torus T, and the sphere S, no degree 1 harmonic map $f : T \to S$ can exist, for if one did, it would have to be holomorphic which is clearly impossible.

Here we survey recent work in Kähler geometry, which uses harmonic maps. The goal in most of this work is to prove an existence result.

This is achieved as a two step process. First one proves (or quotes) an existence theorem for harmonic maps. The second step is to prove that under the hypotheses at hand, all harmonic maps are either holomorphic or conjugate holomorphic.

Frankel Conjecture:

Andreotti and Frankel [7] show that a compact complex manifold of complex dimension 2 admitting a Kähler metric with positive holomorphic bisectional curvature was biholomorphic to $\mathbb{CP}^2$. Frankel conjectured this to be true in arbitrary dimension. Mabuchi [15] showed that Frankel's conjecture is true in dimension 3. As our first application of the method of harmonic maps, we will outline the proof of Frankel's conjecture, due to Siu and Yau [22], in arbitrary dimension.

Theorem. Every compact connected Kähler manifold M^n of positive holomorphic bisectional curvature is biholomorphic to $\mathbb{CP}^n$.

The proof uses two partial results.

Theorem (Kobayashi-Ochiai [11]). $\mathbb{CP}^n$ is characterized, among compact n-dimensional complex manifolds, by the property that its first Chern class $= \lambda c_1(F)$ for some $\lambda \geq n+1$ and some positive holomorphic line bundle F.

Theorem (Bishop-Goldberg [2]). If M is a compact Kähler manifold of positive holomorphic bisectional curvature, then its second Betti number is 1.

We now outline the proof of the Frankel conjecture. There is no loss in assuming that M is simply connected for curvature condition assures that the universal cover $\tilde{M}$ of M is compact and $\mathbb{CP}^n$ has no fixed point free automorphisms. Thus by the Bishop-Goldberg result and the universal coefficient theorem we can conclude that $H^2(M, \mathbb{Z}) = \mathbb{Z}$. Thus, there exists a positive holomorphic line bundle $F \to M$ such that $c_1(F)$ generates $H^2(M, \mathbb{Z})$. Choose a generator g of the free part of $H_2(M, \mathbb{Z})$ satisfying $c_1(F)(g) = 1$. Since M is simply connected the Hurewicz map gives an isomorphism between $\pi_2(M)$ and $H_2(M, \mathbb{Z})$. Let $f : \mathbb{CP}^1 \to M$ represent a homotopy class corresponding to g, under the Hurewicz isomorphism.

We have thus produced a map between $\mathbb{CP}^1$ and M, representing a generator of the free part of $H_2(M, \mathbb{Z})$. Suppose this map f is known to be holomorphic. Let E denote the line bundle corresponding to the divisor of df. Then E is a positive line bundle over $\mathbb{CP}^1$ and $T\mathbb{CP}^1 \otimes E \hookrightarrow f^*TM$. The quotient bundle $\frac{f^*TM}{T\mathbb{CP}^1 \otimes E}$ is a positive bundle over a rational curve. According to a theorem of Grothendieck [8], the quotient splits as a sum of positive line bundles Q_i, $1 \leq i \leq n-1$. Computing Chern classes we find that

$$\begin{aligned} c_1(f^*TM) &= c_1(T\mathbb{CP}^1) + c_1(E) + \sum_{i=1}^{n-1} c_1(Q_i) \\ &\geq c_1(T\mathbb{CP}^1) + (n-1) = n+1 \end{aligned}$$

since $c_1(T\mathbb{CP}^1) = 2$.

We are thus left with the task of proving that f is holomorphic.

Let $f : S^2 \to M$ be a C^1 map with energy $E(f)$, and let $E([f])$ denote the infimum of the sum of the energies of maps whose (homotopy theoretic) sum is homotopic to f. Using the methods of Sacks-Uhlenbeck [17] the following result is obtained.

Theorem: For every C^1 map $f : S^2 \to M$ there exist energy minimizing maps $f_i : S^2 \to M$, $1 \le i \le m$ such that $\sum_{i=1}^{m} f_i$ is homotopic to f and $E([f]) = \sum_{i=1}^{m} E(f_i)$.

The next step in the proof is to show that under certain conditions the maps constructed above are either holomorphic or conjugate holomorphic.

Theorem: If M is a compact Kähler manifold of positive holomorphic bisectional curvature and $f : \mathbb{CP}^1 \to M$ is an energy minimizing map such that $f^*c_1(M)[\mathbb{P}^1] \ge 0$ (resp. ≤ 0) then f is holomorphic (resp. conjugate holomorphic).

This is proved by considering a one complex parameter variation f_t of f and using the second variation formula to compute

$$\frac{\partial^2}{\partial t \partial \bar{t}} \int_{\mathbb{P}^1} |\bar{\partial} f|^2 .$$

This expression is non-negative, by the fact that f is energy minimizing. The curvature condition then guarantees that f is holomorphic.

Hence to prove the Frankel conjecture we need only prove that in the first result $m = 1$, since $c_1(TM)$ is a positive multiple of $c_1(F)$ at least one f_i is holomorphic. To do this, Siu and Yau use holomorphic deformation of rational curves in M. Specifically they show that if $m > 1$, one can holomorphically deform the image of some holomorphic f_i and some conjugate holomorphic f_i so that they are tangent at some point.

By removing a disc centered at the point of tangency and replacing it with a suitable surface one can decrease the energy. Thus $m=1$ and the theorem is proved.

Siu [21] has used the ideas of the proof of the Frankel conjecture to provide a characterization of hyperquadrics. Suppose M is a compact Kähler manifold of non-negative holomorphic bisectional curvature. M is called m-positive at $p \in M$ if for all $v \neq 0 \in T_pM$, $\{w \in T_pM \mid R(v,\bar{v},w,\bar{w}) = 0\}$ is of dimension $< m$.

Theorem: Let M be a compact Kähler manifold of dimension $n \geq 3$. Suppose that M is m-positive everywhere for some $m < \frac{n}{2}+1$ and is 2-positive somewhere on M. Then M is biholomorphic to either complex projective space or the complex hyperquadric.

Strong Rigidity

Calabi and Vesentini [3] showed that compact quotients of bounded symmetric domains are infinitesimally rigid, in the sense that they admit no non-trivial infinitesimal holomorphic deformations. Mostow [16] proved, as a corollary of his work on the strong rigidity of locally symmetric Riemannian manifolds, that in complex dimension at least 2 compact quotients of the ball with isomorphic fundamental groups are either holomorphic or conjugate holomorphic.

In recent work [19], which are described below, Siu has used the method of harmonic maps to extend this result of Mostow. To do this, he introduces the concept of strong negative curvature. If M is a complex manifold with Kähler metric $= 2\mathrm{Re}\, g_{\alpha\bar{\beta}} dz^{\alpha} d\bar{z}^{\beta}$, then the curvature tensor is given by

$$R_{\alpha\bar{\beta}\gamma\bar{\delta}} = \partial_\gamma \partial_{\bar{\delta}} g_{\alpha\bar{\beta}} - g^{\lambda\bar{\mu}} \partial_\alpha g_{\gamma\bar{\mu}} \partial_{\bar{\beta}} g_{\lambda\bar{\delta}}$$

Definition: The curvature is said to be strongly negative if

$$R_{\alpha\bar{\beta}\gamma\bar{\delta}} (A^\alpha \overline{B^\beta} - C^\alpha \overline{D^\beta}) \overline{(A^\delta \overline{B^\gamma} - C^\delta \overline{D^\gamma})} > 0$$

for arbitrary complex $A^\alpha, B^\alpha, C^\alpha, D^\alpha$ whenever $A^\alpha \overline{B^\beta} - C^\alpha \overline{D^\beta} \neq 0$ for some pair (α,β).

If M is strongly negatively curved, then all sectional curvatures are negative, as well as the holomorphic bisectional curvature. The following theorem extends the Mostow rigidity theorem.

Theorem: Let M be a compact Kähler manifold of complex dimension at least 2 which is strongly negatively curved. Suppose N is a compact Kähler manifold which is homotopy equivalent to M. Then N is either biholomorphic or conjugate biholomorphic to M.

We should remark, before outlining the proof of this theorem, that Siu actually proves a stronger result. The curvature tensor of compact quotients of the classical bounded symmetric domains is not strongly negative. Nevertheless, the rigidity theorem above applies to them. We will indicate the necessary modifications in the proof as we proceed.

The other feature of this theorem which should be pointed out is that there is no curvature restriction on N. This is rather surprising in view of Mostow's theorem which requires the competing manifold to be negatively curved. The only requirement on N is that it be Kähler. It is unknown whether this is necessary.

The proof of this result again follows the same two step process. In this case the existence theorem which one needs is the Eells-Sampson theorem 5 which permits one to assume that the homotopy equivalence between M and N is realized by a harmonic map. That this is sufficient is embodied in the following theorem of Siu which is the second step in the proof.

Theorem: Suppose M and N are compact Kähler manifolds with M strongly negatively curved. Suppose $f : N \to M$ is a harmonic map such that $\mathrm{rank}_{\mathbb{R}} df(x) \geq 4$ for some $x \in N$. Then f is either holomorphic or conjugate holomorphic.

To prove the strong rigidity of compact quotients of the classical bounded symmetric domains, Siu shows that if one would allow oneself to strengthen the hypothesis on f_i by requiring it to be a submersion for some $x \in N$, then one can allow some indefiniteness in the curvature tensor of M. We refer the reader to Siu's paper for the precise statement.

To prove the complex analyticity of the map f, we only require that f have rank at least 4 in an open set. This is because it is sufficient to prove that f is holomorphic on an open set. Once this is proved, Aronszajn's unique continuation principle [1] can be applied to conclude that f is holomorphic on N.

The proof is via a Bochner type argument. The difficulty here is that if one considers the Laplacian of $|\bar{\partial}f|^2$ curvature terms from both

M and N will appear. Further, the curvatures will appear with opposite signs. What Siu does is replace $|\bar{\partial}f|^2$ by $\bar{\partial}f \wedge \overline{\bar{\partial}f} \wedge \omega^{n-2}$, where ω is the Kähler form on N; and replace the Laplacian by $\partial\bar{\partial}$. In this way he is able to get rid of terms involving the curvature tensor on N.

Deformation Theory

Siu's theorem has been used in [9] to study the deformation theory of complex submanifolds of strongly negatively curved compact Kähler manifolds, answering a question of Siu [20].

Suppose $N_0 \xrightarrow{f_0} M$ is a holomorphic imbedding of a compact Kähler manifold of complex dimension at least 2 into a strongly negatively curved Kähler manifold M. Associated with this situation there are two deformation spaces which we now describe. If we neglect the imbedding and consider N_0 merely as a compact complex manifold, then there is the versal deformation $\mathcal{N}_k \to (T_k, 0)$ centered at N_0 of the complex structure on N_0. According to [14], this is a holomorphic family which, up to obvious equivalence, contains every family of complex structures centered at N_0. This is also the universal deformation of N_0 as a submanifold of M. According to [4] there is a unique holomorphic family $\mathcal{N}_d \to (T_d, 0)$ which parametrizes deformations of N_0 inside of M.

Theorem. If M is a strongly negatively curved compact Kähler manifold and N_0 is a Kähler submanifold of complex dimension at least 2, then the families $\mathcal{N}_d \to T_d$ and $\mathcal{N}_k \to T_k$ coincide.

The idea of the proof is straightforward. If N_t is a deformation of N_0, depending smoothly on t, then for $|t|$ small, N_t is Kähler [13]. We now use the Eells-Sampson theorem or find a family of imbeddings

$f_t : N_t \to M$; which, according to [18], depend smoothly on t. According to Siu's theorem, each f_t is either holomorphic or conjugate holomorphic. Since f_0 is holomorphic, it is easy to see that each f_t is (for $|t|$ small). Thus, in order to prove our theorem, we need to show that $t \mapsto f_t$ is holomorphic. The way this is done is to show that $\frac{\partial f_t}{\partial \bar{t}}$ is holomorphic as a section of $f_t^{-1}TM \to N_t$. It is a consequence of the negativity of the holomorphic bisectional curvature on M, and the holomorphicity of f_t, that the bundle $f_t^{-1}TM$ admits no non-zero holomorphic sections. For this see [12].

We remark that the idea of taking a parametrized family of holomorphic maps and differentiating it with respect to the parameter to obtain a holomorphic section of a bundle, which is then shown to be zero by imposing conditions, has been used in [10] to obtain finiteness and rigidity theorems for spaces of holomorphic maps. For example, it can be shown that if M is compact and has negative holomorphic bisectional curvature (with no Kähler assumption) and N is a compact complex space, then there are only a finite number of non-constant holomorphic maps $f : N \to M$.

M. Kalka
Department of Mathematics
Tulane University
New Orleans, La. 70118

Research partially supported by N.S.F. Grant

Bibliography

1. N. Aronszajn, "A unique continuation theorem for solutions of elliptic partial differential equations or inequalities of second order," J. Math. Pures Appl. 36 (1957), p.235-249.

2. R. L. Bishop and S. I. Goldberg, "On the second cohomology group of a Kähler manifold of positive curvature," Proc. Amer. Math. Soc. 16 (1965), p.119-122.

3. E. Calabi and E. Vesentini, "On compact locally symmetric Kahler manifolds," Ann. of Math. 71 (1960), p.472-507.

4. A. Douady, Le problème des modules pour les sous-espaces analytiques d'un espace analytique donné, Ann. Inst. Fourier (Grenoble) 16 (1966), p.1-95.

5. J. Eells and J. H. Sampson, Harmonic mappings of Riemannian manifolds," Amer. J. Math. 86 (1964), p.109-160.

6. J. Eells and J. C. Wood, "Restrictions on harmonic maps of surfaces," Topology 15 (1976), p.263-266.

7. T. Frankel, "Manifolds with positive curvature," Pacific J. Math. 11 (1961), p.165-174.

8. A. Grothendieck, "Sur la classification des fibres holomorphes sur la sphère de Riemann," Amer. J. Math. 79 (1957), p.121-138.

9. M. Kalka, "Deformation of submanifolds of strongly negatively curved manifolds," Math. Ann. 251 (1980), p.243-248.

10. M. Kalka, B. Shiffman and B. Wong, "Finiteness and rigidity theorems for holomorphic mappings," to appear.

11. S. Kobayashi and T. Ochiai, "Characterizations of complex projective space and hyperquadrics," J. Math Kyoto Univ. 13 (1973), p.31-47.

12. S. Kobayashi and J. Wu, "On holomorphic sections of certain hermitian vector bundles," Math. Ann. 189 (1970), p.1-4.

13. K. Kodaira and D. C. Spencer, "On deformations of complex analytic structures, III, Stability theorems," Ann. of Math. 71 (1960), p.43-76.

14. M. Kuranishi, "On locally complete families of complex analytic structures," Ann. of Math. 75 (1962), p.536-577.

15. T. Mabuchi, "$\mathbb{C}$-actions and algebraic threefolds with ample tangent bundle," Nagoya Math. J. 69 (1978), p.33-64.

16. G. D. Mostow, Strong Rigidity of Locally Symmetric Spaces, Ann. of Math. Studies 78 (1973), Princeton University Press.

17. J. Sacks and K. Uhlenbeck, "The existence of minimal immersions of 2-spheres", preprint.

18. J. H. Sampson, "Some properties and applications of harmonic mappings," Ann. Sci. École Norm. Sup. 11 (1978), p.211-228.

19. Y. T. Siu, "The complex-analyticity of harmonic maps and the strong rigidity of compact Kahler manifolds," Ann. of Math. 112 (1980), p.73-111.

20. ________, "Some remarks on the complex analyticity of harmonic maps," preprint.

21. ________, "Curvature characterization of hyperquadrics," Duke Math. J. 47 (1980), 641-654.

HARMONIC FOLIATIONS *)

by
Franz W. Kamber and Philippe Tondeur

INTRODUCTION.

Let F be a codimension q foliation on a manifold M. All data are assumed to be smooth. Recently Rummler [RU1], [RU2], Sullivan [S] and Haefliger [HA] discussed the question of the existence of a Riemannian metric on M for which all the leaves of F are minimal submanifolds. A foliation admitting such a metric is called geometrically taut.

In this paper we consider an arbitrary Riemannian metric on M. The orthogonal splitting $TM \cong L \oplus L^{\perp}$ defined by the foliation $L \subset TM$ and the metric identifies $L^{\perp}$ with the normal bundle $Q = TM/L$, and thus induces a metric on Q. The Riemannian connection on M further induces a connection ∇ on Q. These data allow to define on the Q-valued differential forms on M operators d_∇, d_∇^* and a Laplace operator in the usual fashion (see section 3). The projection $\pi : TM \to Q$ can be viewed as a 1-form on M with values in Q. We prove the following result.

THEOREM. Let F be a foliation on a manifold M and g_M a Riemannian metric. Then all the leaves of the foliation are minimal submanifolds of M if and only if the canonical Q-valued 1-form $\pi : TM \to Q$ is harmonic.

The proof is very much in the spirit of Eells and Sampson's theory of harmonic maps [ES]. For a foliation defined by a Riemannian submersion $f : M \to N$ it is indeed a result of theirs that the map f is harmonic if and only if the leaves are minimal submanifolds of M. Our observation is that several geometric objects interesting for the discussion of submersions can be attached to foliations, since these are locally defined by submersions. This is in particular true of the second fundamental form and its trace, the mean curvature. The latter is up to a normalization the tension field of the foliation defined in 2.24. The

---

*) Research partially supported by NSF Grant MCS 79-00256.

vanishing of the second fundamental form of the foliation implies that all leaves are totally geodesic submanifolds. The tension field equal $d_\nabla^* \pi$ (Proposition 3.2). This implies that the harmonicity of π is equivalent to the vanishing of the tension field, which in turn expres the fact that all leaves are minimal (Theorems 2.28 and 3.3). Example of harmonic foliations are discussed at the end of section 3.

The energy of a foliation F on a compact oriented manifold M defined in terms of $\pi : TM \to Q$ by

$$E(F) = \tfrac{1}{2} \|\pi\|^2.$$

See section 4 for more details. For a foliation defined by a Riemanni submersion $f : M \to N$ this is the Eells-Sampson definition for the energy of the submersion [ES]. One expects therefore a characterizati of harmonic Riemannian foliations as extremals of the energy functiona under appropriate variations. We prove the following result.

THEOREM. <u>Let</u> M <u>be a compact oriented manifold equipped with a Riemannian foliation</u> F <u>and a bundle-like metric. Then</u> F <u>is harmor if and only if it is an extremal of the energy functional</u> E <u>under variations through Riemannian foliations with fixed normal bundle and holonomy on</u> M.

Several statements in this introduction call for clarifications c terminology. We begin with that task.

1. FOLIATIONS AND METRICS

Let $L \subset TM$ be a foliation with normal bundle $Q = TM/L$, and g a Riemannian metric on M. The metric g_M defines a splitting σ in the exact sequence

$$(1.1) \qquad 0 \longrightarrow L \longrightarrow TM \underset{\pi}{\overset{\sigma}{\rightleftarrows}} Q \longrightarrow 0$$

with $\sigma Q = L^\perp$ the orthogonal complement of L. Thus g_M induces a metric g_Q on Q.

Let $\overset{\circ}{\nabla}$ denote the (partial) Bott connection in Q defined by

$$(1.2) \qquad \overset{\circ}{\nabla}_X s = \pi[X,Y] \quad \text{for} \quad X \in \Gamma L, \quad Y \in \Gamma TM \quad \text{with} \quad \pi(Y) = s.$$

Let ∇^M denote the Riemannian connection of g_M. A connection ∇ in Q is then defined by

$$(1.3)\quad \begin{cases} \nabla_X s = \pi[X,Y_s] & \text{for } X \in \Gamma L,\ s \in \Gamma Q \text{ and } Y_s = \sigma(s) \in \Gamma\sigma Q \\ \nabla_X s = \pi(\nabla^M_X Y_s) & \text{for } X \in \Gamma\sigma Q,\ s \in \Gamma Q \text{ and } Y_s = \sigma(s) \in \Gamma\sigma Q. \end{cases}$$

The first condition says that ∇ is an adapted connection on Q, i.e. a connection extending the partial Bott connection along L (see e.g. [KT3] for the terminology used).

Let ∇ be any connection in the normal bundle Q of a foliation. Its torsion is the Q-valued 2-form on M defined by

$$(1.4)\qquad T_\nabla(X,Y) = \nabla_X\pi(Y) - \nabla_Y\pi(X) - \pi[X,Y]$$

for $X,Y \in \Gamma TM$.

1.5 LEMMA. <u>The connection</u> ∇ <u>in</u> Q <u>defined by</u> (1.3) <u>is torsion free.</u>

Proof. For $X \in \Gamma L$, $Y \in \Gamma TM$ we have by (1.2) $T_\nabla(X,Y) = 0$. It further follows by (1.3) that

$$T_\nabla(X,Y) = \pi T_{\nabla^M}(X,Y) \quad \text{for } X,Y \in \Gamma\sigma Q.$$

But ∇^M is torsion free, i.e. $T_{\nabla^M} = 0$. It follows that $T_\nabla = 0$. ∎

A connection ∇ in Q is metric with respect to a metric g_Q on Q if $\nabla_X g_Q = 0$ for $X \in \Gamma TM$, i.e.

$$(1.6)\qquad Xg_Q(s,t) = g_Q(\nabla_X s,t) + g_Q(s,\nabla_X t)$$

for all $s,t \in \Gamma Q$. This need not be the case for the connection ∇ defined by (1.3) and g_Q induced by a metric g_M on M. This condition is in fact not necessary for part of our characterizations of harmonicity, and we will precisely state where the metric properties of ∇ are required.

For clarity's sake let us review the following concepts due to Reinhart [RE1]. A foliation is Riemannian (a R-foliation), if the normal bundle is equipped with a holonomy invariant fiber metric g_Q. This condition is expressed in terms of the Bott connection $\overset{\circ}{\nabla}$ by $\overset{\circ}{\nabla}_X g_Q = 0$ for $X \in \Gamma L$, i.e. formula (1.6) with the proviso $X \in \Gamma L$.

A Riemannian metric g_M on M is bundle-like with respect to the foliation $L \subset TM$, if the fiber metric g_Q induced on Q turns the foliation into a R-foliation. A R-foliation admits a bundle-like metric: choose any fiber metric g_L on L, a splitting σ of (1.1), and set g_M equal to the orthogonal sum $g_L \oplus g_Q$ on $TM \cong L \oplus \sigma Q$.

For later use we note that if the foliation F is deformed throu R-foliations $F_t (L_t \subset TM$ for $t \in (-\varepsilon,\varepsilon)$, $F_0 = F)$, a metric g_M on which is bundle-like for $L_0 = L$ is not necessarily so for the foliations L_t, $t \neq 0$. The condition stated in the second theorem of the introduc tion implies that $Q_t = TM/L_t$ is independent of t, i.e. equal Q_0 = (The additional requirement of the t-independence of the holonomy wil be made clear in section 4, when we discuss variations of F associat to sections of Q.)

Returning now to metric properties of any connection ∇ on the normal bundle Q of a foliation, we prove the following result.

1.7 PROPOSITION. Let $L \subset TM$ be a foliation with normal bundle Q, and g_Q any fiber metric on Q. If ∇ is a metric and torsionfree connection in Q, then

$$
\begin{aligned}
2g_Q(\nabla_X s,t) &= Xg_Q(s,t) + Y_s g_Q(\pi(X),t) - Y_t g_Q(\pi(X),s) \\
&\quad + g_Q(\pi[X,Y_s],t) + g_Q(\pi[Y_t,X],s) - g_Q(\pi[Y_s,Y_t],\pi(X))
\end{aligned}
\tag{1.8}
$$

for $X \in \Gamma TM$; $s,t \in \Gamma Q$; $Y_s, Y_t \in \Gamma TM$ with $\pi(Y_s) = s$, $\pi(Y_t) = t$.

Proof. Expanding the first three terms on the RHS of (1.8) using (1 and the definition (1.4) with $T_\nabla = 0$, yields (1.8) by a straightforwa calculation. ∎

1.9 REMARK. For the point foliation we have $L = 0$ and $Q = TM$. Th formula (1.8) is then the classical formula which determines the Riemannian connection on M by the RHS ([KN, Vol. I, p. 160]).

1.10 COROLLARY. For a given fiber metric g_Q on Q there is at mos one metric and torsionfree connection in Q.

This result is of interest for R-foliations as follows.

1.11 THEOREM. (i) Let g_M be a bundle-like metric on the foliated manifold M. Then the connection ∇ on Q defined by (1.3) is tors free and metric.

(ii) Let $L \subset TM$ be a R-foliation with holonomy invariant fib metric g_Q on the normal bundle Q. There is a unique metric and torsionfree connection on Q.

Proof. (i) ∇ is torsionfree by Lemma 1.5. The metric condition (1 holds for $X \in \Gamma L$ by assumption. For $X \in \Gamma Q$ and $s,t \in \Gamma Q$ we ha

$$Xg_Q(s,t) = Xg_M(Y_s,Y_t) \quad \text{for} \quad Y_s = \sigma(s),\ Y_t = \sigma(t).$$

Using successively that ∇^M is metric, sum $L \oplus \sigma Q$ orthogonal, σ a bundle isometry, and the definition (1.3), we find that

$$\begin{aligned} Xg_M(Y_s,Y_t) &= g_M(\nabla^M_X Y_s, Y_t) + g_M(Y_s, \nabla^M_X Y_t) \\ &= g_M(\sigma\pi(\nabla^M_X Y_s),\ \sigma(t)) + g_M(\sigma(s), \sigma\pi(\nabla^M_X Y_t)) \\ &= g_Q(\pi(\nabla^M_X Y_s), t) + g_Q(s, \pi(\nabla^M_X Y_t)) \\ &= g_Q(\nabla_X s, t) + g_Q(s, \nabla_X t) \end{aligned}$$

which proves that ∇ is indeed metric. (ii) follows from (i) combined with the uniqueness result 1.10. ∎

1.12 REMARKS. Pasternack [P] was the first to point out that the existence of the connection given by (1.3) results in an improvement of Bott's vanishing theorem by a factor two for R-foliations. Note that since the foliation is locally defined by submersions related via Riemannian isometries in the target space, the Riemannian connection in the target space pulls back to the connection ∇ in Q. From this description of ∇ it follows that it is a basic connection [KT3], which means that it satisfies

(1.13) $$i(X)R_\nabla = 0 \quad \text{for} \quad X \in \Gamma L,$$

where R_∇ is the curvature of ∇ considered as an End Q-valued 2-form on M.

We return now to the initial situation of a foliated manifold M and an arbitrary Riemannian metric g_M on M. A connection ∇' in Q can be defined by the formula

(1.14) $$\nabla'_X s = \pi(\nabla^M_X Y_s) \quad \text{for} \quad X \in \Gamma TM,\ s \in \Gamma Q \ \text{and} \ Y_s = \sigma(s) \in \Gamma\sigma Q.$$

This coincides with (1.3) for $X \in \Gamma\sigma Q$, but not necessarily for $X \in \Gamma L$.

1.15 LEMMA. <u>The connection ∇' on Q is metric (with respect to the induced metric g_Q on Q).</u>

Proof. This follows by exactly the same calculation as in the proof of part (i) of Theorem 1.11. ∎

1.16 LEMMA. <u>For the torsion $T_{\nabla'}$ of ∇' we have</u>

(i) $T_{\nabla'}(X,Y) = 0$ <u>for</u> $X,Y \in \Gamma L$;

(ii) $T_{\nabla'}(X,Y) = 0$ <u>for</u> $X,Y \in \Gamma\sigma Q$;

(iii) $T_{\nabla'}(X,Y) = \pi(\nabla^M_Y X)$ <u>for</u> $X \in \Gamma L,\ Y \in \Gamma\sigma Q$.

Proof. (i) is obvious from (1.4). (ii) follows because ∇^M is tor-sionfree. (iii) follows for $X \in \Gamma L, Y_s = \sigma(s)$ from

$$T_{\nabla'}(X,Y_s) = \nabla'_X s - \pi[X,Y_s] = \pi(\nabla^M_X Y_s - [X,Y_s]) = \pi(\nabla^M_{Y_s} X)$$

since ∇^M is torsionfree. ∎

Note that the last calculation shows in fact that

$$(1.17)\quad T_{\nabla'}(X,Y_s) = \nabla'_X s - \overset{\circ}{\nabla}_X s = \pi(\nabla^M_{Y_s} X) \quad \text{for} \quad X \in \Gamma L,\ s \in \Gamma Q.$$

1.18 LEMMA. ∇' <u>is determined by the formula</u>

$$2g_Q(\nabla'_X s,t) = Xg_Q(s,t) + Y_s g_Q(\pi(X),t) - Y_t g_Q(\pi(X),s)$$
$$+ g_Q(\pi[X,Y_s],t) + g_Q(\pi[Y_t,X],s) - g_M(\pi[Y_s,Y_t],X)$$

<u>for</u> $X \in \Gamma TM;\ s,t \in \Gamma Q;\ Y_s = \sigma(s),\ Y_t = \sigma(t) \in \Gamma\sigma Q$.

Proof. By definition of ∇' and the orthogonality of σ clearly

$$2g_Q(\nabla'_X s,t) = 2g_Q(\pi(\nabla^M_X Y_s),t) = 2g_M(\nabla^M_X Y_s, Y_t).$$

The result follows by applying the classical expansion formula to the last term (the special case of Proposition 1.7 for $L = 0$, $Q = TM$ and $\pi = id$), and using again the orthogonality of σ to replace g_M by g_Q whenever possible (except in the last term). ∎

We can now compare the connections ∇ and ∇' as follows.

1.19 PROPOSITION. <u>Let</u> M <u>be a foliated manifold with an arbitrary Riemannian metric</u> g_M <u>on</u> M. <u>Let</u> g_Q <u>be the induced metric in the normal bundle</u> Q <u>with respect to the orthogonal splitting</u> σ <u>of</u> (1. <u>On</u> Q <u>we have by</u> (1.3) <u>a torsionfree connection</u> ∇, <u>and by</u> (1.15) <u>a metric connection</u> ∇'. <u>The following conditions are equivalent:</u>

(i) $\nabla = \nabla'$:

(ii) $T_{\nabla'} = 0$;

(iii) <u>the subbundle</u> $\sigma Q \subset TM$ <u>is involutive and the metric</u> g_Q <u>holonomy invariant</u>;

(iv) $\nabla^M_{Y_s} Y_t \in \Gamma\sigma Q$ <u>for</u> $s,t \in \Gamma Q;\ Y_s = \sigma(s),\ Y_t = \sigma(t) \in \Gamma\sigma Q$.

Proof. Comparing (1.3) and (1.14) it is clear that $\nabla = \nabla'$ iff

$$(1.20)\qquad \nabla'_X s - \overset{\circ}{\nabla}_X s = 0 \quad \text{for} \quad X \in \Gamma L,\ s \in \Gamma Q.$$

Formula (1.17) and Lemma 1.16 establish the equivalence of (i) and (i

It remains to prove the equivalence of (i), (iii) and (iv). For $X \in \Gamma L$ we get from Lemma 1.18.

$$2g_Q(\nabla'_X s,t) = Xg_Q(s,t) + g_Q(\pi[X,Y_s],t) + g_Q(\pi[Y_t,X],s) - g_M(\pi^{\perp}[Y_s,Y_t],X)$$

where in the last term $\pi^{\perp} : TM \to L$ denotes the orthogonal projection along σQ, i.e. $\pi^{\perp} = \mathrm{id} - \sigma\circ\pi$.

Now

$$g_Q(\pi[X,Y_s],t) = g_Q(\overset{\circ}{\nabla}_X s,t)$$

$$g_Q(\pi[Y_t,X],s) = -g_Q(s,\overset{\circ}{\nabla}_X t)$$

and

$$(\overset{\circ}{\nabla}_X g_Q)(s,t) = Xg_Q(s,t) - g_Q(\overset{\circ}{\nabla}_X s,t) - g_Q(s,\overset{\circ}{\nabla}_X t).$$

Therefore

$$(1.21) \qquad 2g_Q(\nabla'_X s - \overset{\circ}{\nabla}_X s,t) = (\overset{\circ}{\nabla}_X g_Q)(s,t) - g_M(\pi^{\perp}[Y_s,Y_t],X).$$

From (1.17) and the formula

$$(1.22) \qquad g_Q(\pi(\nabla^M_{Y_s} X),t) = g_M(\nabla^M_{Y_s} X,Y_t) = -g_M(X,\nabla^M_{Y_s} Y_t) + Y_s g_M(X,Y_t) = -g_M(X,\nabla^M_{Y_s} Y_t)$$

we obtain the identities:

$$(1.23) \qquad g_Q(\pi(\nabla^M_{Y_s} X),t) = -g_M(X,\nabla^M_{Y_s} Y_t) = g_Q(\nabla'_X s - \overset{\circ}{\nabla}_X s,t) = \tfrac{1}{2}((\overset{\circ}{\nabla}_X g_Q)(s,t) - g_M(X,\pi^{\perp}[Y_s,Y_t])).$$

The implications (i) $\Leftrightarrow$ (iv) and (iii) $\Rightarrow$ (i) follow directly from this formula. If $\nabla = \nabla'$, then ∇ is a metric connection, hence $\overset{\circ}{\nabla}_X g_Q = \nabla_X g_Q = 0$ for $X \in \Gamma L$ and thus (i) $\Rightarrow$ (iii) by (1.23). ∎

2. SECOND FUNDAMENTAL FORM AND TENSION FIELD.

Let $E \to M$ be a vectorbundle. For a connection ∇ on E we consider the exterior derivative $d_\nabla : \Omega^r(M,E) \to \Omega^{r+1}(M,E)$, $r \geq 0$ on E-valued forms given for $\omega \in \Omega^r(M,E)$ by

$$(d_\nabla\omega)(X_1,\ldots,X_{r+1}) = \sum_{i=1}^{r+1} (-1)^{i+1}\nabla_{X_i}\omega(X_1,\ldots,\hat{X}_i,\ldots,X_{r+1}) + \sum_{i<j} (-1)^{i+j}\omega([X_i,X_j],X_1,\ldots,\hat{X}_i,\ldots,\hat{X}_j,\ldots,X_{r+1})$$

where the X_i's are vectorfields on M. Then $d^2_\nabla \omega = R_\nabla \wedge \omega$ with R_∇ the curvature of ∇.

If M is Riemannian with Riemannian connection ∇^M, the covari derivative $\nabla_X : \Omega^r(M,E) \to \Omega^r(M,E)$, $X \in \Gamma TM$, $r \geq 0$ is for $\omega \in \Omega^r(M,$ given by

$$(\nabla_X\omega)(X_1,\ldots,X_r) = \nabla_X\omega(X_1,\ldots,X_r) - \sum_{i=1}^{r} \omega(X_1,\ldots,\nabla^M_X X_i,\ldots,X_r)$$

where the X_i's are vectorfields on M. By definition

$$(\nabla\omega)(X;X_1,\ldots,X_r) = (\nabla_X\omega)(X_1,\ldots,X_r).$$

Because ∇^M is torsionfree, $d_\nabla\omega$ can be expressed as the alternatio of $\nabla\omega$, i.e.

$$(d_\nabla\omega)(X_1,\ldots,X_{r+1}) = \sum_{i=1}^{r+1} (-1)^{i+1}(\nabla_{X_i}\omega)(X_1,\ldots,\hat{X}_i,\ldots,X_{r+1}).$$

For $r = 1$ this formula reads

(2.1) $(d_\nabla\omega)(X,Y) = (\nabla_X\omega)(Y) - (\nabla_Y\omega)(X) = (\nabla\omega)(X,Y) - (\nabla\omega)(Y,X).$

We apply this to the normal bundle $Q \to M$ of a foliation on M. For a Riemannian metric g_M on M we have then by (1.3) a connectio ∇ on Q. For the bundle projection $\pi : TM \to Q$ viewed as $\pi \in \Omega^1(M$ we have then the following fact.

2.2 PROPOSITION. $d_\nabla\pi = 0$.

Proof. By (1.4) $d_\nabla\pi = T_\nabla$, and by Lemma 1.5 ∇ is torsionfree. ■

REMARK. For any connection ∇ in the normal bundle, the torsion can thus be defined by $T_\nabla = d_\nabla\pi$.

2.3 PROPOSITION. _The Q-valued bilinear form on_ M

$$\alpha(X,Y) = -(\nabla\pi)(X,Y) \equiv \pi(\nabla^M_X Y) - \nabla_X\pi(Y)$$

is symmetric.

Proof. This follows from (2.1) and (2.2). ■

We call α the _second fundamental form_ of the foliation. To justify the terminology, observe that

(2.4) $\alpha(X,Y) = \pi(\nabla^M_X Y)$ for $X \in \Gamma TM$, $Y \in \Gamma L$.

It follows in particular that α restricted to any leaf L of the foliation is the second fundamental form of the submanifold $L \subset M$. For a submersion $f : M \to N$, the second fundamental form as defined above is the second fundamental form of f in the sense of Eells-Sampson [ES] (up to sign).

2.5 REMARK. The property that α restricted to any leaf of the foliation is the second fundamental form of the embedded leaf is common to both the Q-valued symmetric bilinear form T_1 on M defined by Reinhart [RE2, p. 620], and the TM-valued bilinear form T_2 on M defined by O'Neill [ON, p. 460]. But no two of these three forms α, T_1, T_2 coincide on M (but they have further common properties from what transpires below).

2.6 REMARK. $\alpha(X,Y) = 0$ for all $X,Y \in \Gamma L$ if and only if all leaves of the foliation are totally geodesic submanifolds. The condition that α vanishes for all $X,Y \in \Gamma L$ is by (2.4) equivalent to the property that each leaf L is an auto-parallel submanifold of M. In view of the torsionfreeness of ∇^M auto-parallel and totally geodesic submanifolds of M coincide [KN, Vol. II, p. 56,57].

2.7 REMARK. The tensors T_1, T_2 of Reinhart and O'Neill mentioned above do vanish entirely in this situation. But this need not be so for α.

For the further discussion of α it is convenient to introduce for each $\nu \in \Gamma Q$ the map $A(\nu) : TM \to TM$ uniquely defined by

$$g_M(A(\nu)X,Y) = g_Q(\alpha(X,Y),\nu) \quad \text{for} \quad X,Y \in \Gamma TM. \tag{2.8}$$

The symmetry of α implies the selfadjointness of $A(\nu)$. With respect to the decomposition (1.1) the map $A = A(\nu)$ has then the matrix representation

$$A = \begin{pmatrix} A_1 & A_2 \\ A_2^* & A_3 \end{pmatrix} \tag{2.9}$$

with $A_1 : L \to L$, $A_3 : \sigma Q \to \sigma Q$ both selfadjoint and $A_2 : \sigma Q \to L$.

We consider further the Weingarten map $W(\nu) : L \to L$ given for $\nu \in \Gamma Q$ by

$$W(\nu)X = -(\pi^{\perp})(\nabla^M_X Y_\nu) \quad \text{for} \quad X \in \Gamma L, Y_\nu = \sigma(\nu) \in \Gamma\sigma Q. \tag{2.10}$$

Here $\pi^{\perp} = id - \sigma\circ\pi : TM \to L$ denotes as before the orthogonal projection along $\sigma Q \subset TM$.

2.11 LEMMA. $A_1(\nu) = W(\nu)$.

Proof. Let $X,Y \subset \Gamma L$. Then for fixed $\nu \in \Gamma Q$ we obtain from (2.4) and the orthogonality of σ

$$g_M(A_1(\nu)X,Y) = g_Q(\alpha(X,Y),\nu) = g_Q(\pi(\nabla^M_X Y),\nu) = g_M(\nabla^M_X Y, Y_\nu)$$
$$= Xg_M(Y,Y_\nu) - g_M(Y,\nabla^M_X Y_\nu) = g_M(Y,W(\nu)X).$$

The desired result follows. ∎

2.12 LEMMA. $A_3 = 0$.

Proof. Let $s,t \in \Gamma Q$ and $Y_s = \sigma(s)$, $Y_t = \sigma(t) \in \Gamma\sigma Q$.

Then for fixed $\nu \in \Gamma Q$

$$g_M(A_3(\nu)Y_s,Y_t) = g_Q(\alpha(Y_s,Y_t),\nu) = g_Q(\pi(\nabla^M_{Y_s} Y_t) - \nabla_{Y_s} Y_t,\nu).$$

But by (1.3) the last term vanishes. Thus $A_3(\nu)Y_s = 0$ for all $s \in \Gamma Q$ and therefore $A_3 = 0$. ∎

Next we evaluate $A_2(\nu) : \sigma Q \to L$.

2.13 LEMMA. <u>For</u> $X \in \Gamma L$ <u>and</u> $\nu,s \in \Gamma Q$

$$2g_M(A_2(\nu)Y_s,X) = (\overset{\circ}{\nabla}_X g_Q)(s,\nu) - g_M(\pi^\perp[Y_s,Y_\nu],X),$$

<u>where</u> $Y_s = \sigma(s)$, $Y_\nu = \sigma(\nu)$.

<u>If in particular</u> g_Q <u>is holonomy invariant</u>, <u>i.e.</u> $\overset{\circ}{\nabla}_X g_Q = 0$, <u>then</u>

$$A_2(\nu)Y_s = -\tfrac{1}{2}\,\pi^\perp[Y_s,Y_\nu].$$

Proof. Let $X \in \Gamma L$, $\nu,s \in \Gamma Q$. Then we obtain from (2.4)

$$g_M(A_2(\nu)Y_s,X) = g_Q(\alpha(Y_s,X),\nu)$$
$$= g_Q(\pi(\nabla^M_{Y_s} X),\nu)$$

and from (1.23)

$$g_M(A_2(\nu)Y_s,X) = -g_M(\pi^\perp(\nabla^M_{Y_s} Y_\nu),X)$$
$$= \tfrac{1}{2}((\overset{\circ}{\nabla}_X g_Q)(s,\nu) - g_M(\pi^\perp[Y_s,Y_\nu],X)). \quad ∎$$

If we use again the metric connection ∇' defined by (1.14), then by (1.23) we have further

$$(2.14)\qquad \begin{aligned} g_M(A_2(\nu)Y_s,X) &= g_Q(\nabla'_X s - \overset{\circ}{\nabla}_X,\nu) \quad \text{and} \\ \alpha(Y_s,X) &= \nabla'_X s - \overset{\circ}{\nabla}_X s\,. \end{aligned}$$

The first part of the next Proposition summarizes the preceding discussion.

2.15 PROPOSITION. Let M be a foliated manifold, and g_M an arbitrary Riemannian metric on M. With respect to the corresponding splitting of (1.1) the operator $A(\nu) : TM \to TM$ defined by (2.8) for every $\nu \in \Gamma Q$ has the matrix representation

$$(2.16)\qquad A(\nu) = \begin{pmatrix} W(\nu) & A_2(\nu) \\ A_2^*(\nu) & 0 \end{pmatrix}$$

where $W(\nu) : L \to L$ is the Weingarten map (2.10), and $A_2(\nu) : \sigma Q \to L$ is given by (2.13) or equivalently (2.14).

Further, the following conditions are equivalent:

(i) $A_2 = 0$;

(ii) the subbundle $\sigma Q \subset TM$ is involutive and the metric g_Q is holonomy invariant (i.e. F is a R-foliation);

(iii) $\pi^\perp(\nabla^M_{Y_s} Y_t) = 0$ for $s,t \in \Gamma Q$, $Y_s = \sigma(s)$, $Y_t = \sigma(t)$, i.e. the distribution $\sigma Q \subset TM$ is totally geodesic.

Proof. This follows from Lemmas 2.11 to 2.13 and Proposition 1.19. For the second statement in (iii) one has to observe that the term $\pi^\perp(\nabla^M_{Y_s} Y_t)$ is by (2.4) equal to the second fundamental form $\alpha^\perp(Y_s,Y_t)$ of the complementary distribution $\sigma Q \subset TM$. Its vanishing for all $s,t \in \Gamma Q$ implies $\pi^\perp[Y_s,Y_t] = 0$, that is the involutivity of $\sigma Q \subset TM$. ∎

Part of the last statements are proved by Rummler [RU1]. Together with (2.6), one gets the following consequence.

2.17 COROLLARY. The following conditions on the foliation F are equivalent:

(i) $\alpha = 0$;

(ii) the leaves of F are totally geodesic, the foliation is Riemannian, and the distribution $\sigma Q \subset TM$ is involutive;

(iii) the distributions $L \subset TM$ and $\sigma Q \subset TM$ are both totally geodesic.

Note that for a codimension one foliation the bundle Q is necessarily involutive, since it is a line bundle. Condition (ii) says then that F is a R-foliation with totally geodesic leaves.

The following evaluation formulas for α will be useful. Let $E_1,\dots,E_n$ be a local orthonormal frame of TM with $E_i \in \Gamma L$ for $i = 1,\dots,p$ and $E_\beta \in \Gamma\sigma Q$ for $\beta = p+1,\dots,n (p = \dim L,\ n = \dim M)$.

2.18 LEMMA. <u>For</u> $1 \leq i, j \leq p$

$$\alpha(E_i,E_j)_\beta = -\tfrac{1}{2}([E_i,E_\beta]_j + [E_j,E_\beta]_i),$$

<u>where the subscript denotes the component in the direction of the corresponding vectorfield.</u>

Proof. This results from the following calculation

$$2\alpha(E_i,E_j)_\beta = 2g_Q(\alpha(E_i,E_j),E_\beta) = 2g_Q(\pi(\nabla^M_{E_i}E_j),E_\beta) = 2g_M(\nabla^M_{E_i}E_j,E_\beta)$$
$$= g_M([E_i,E_j],E_\beta) + g_M([E_\beta,E_i],E_j) - g_M([E_j,E_\beta],E_i).$$

The first term vanishes, and the result follows. ∎

2.19 COROLLARY.

(i) $$\alpha(E_i,E_j) = -\tfrac{1}{2}\sum_{\beta=p+1}^{n}\{[E_i,E_\beta]_j + [E_j,E_\beta]_i E_\beta \in \Gamma Q$$

(ii) $$W(E_\beta)E_i = -\tfrac{1}{2}\sum_{j=1}^{p}\{[E_i,E_\beta]_j + [E_j,E_\beta]_i\}E_j \in \Gamma L.$$

For the following calculations we assume that F is a R-foliation and g_M a bundle-like metric on M. The effect is that the term $\overset{\circ}{\nabla}_X g$ in (2.13) disappears. With the same notations as above we have the following result.

2.20 PROPOSITION. <u>For</u> $p+1 \leq \beta, \gamma \leq n$

$$A_2(E_\beta)E_\gamma = \tfrac{1}{2}\sum_{i=1}^{p}[E_\beta,E_\gamma]_i E_i \in \Gamma L.$$

Proof. This follows from the formula in 2.13. ∎

2.21 COROLLARY. <u>For</u> $1 \leq i \leq p,\ p+1 \leq \gamma \leq n$

$$\alpha(E_\gamma,E_i) = \tfrac{1}{2}\sum_{\beta=p+1}^{n}[E_\beta,E_\gamma]_i E_\beta .$$

Proof. This follows from 2.20 and

$$g_M(A_2(E_\beta)E_\gamma,E_i) = g_M(\alpha(E_\gamma,E_i),E_\beta).$$ ∎

From 2.20 we have immediately

$$(2.22)\qquad A_2^*(E_\beta)E_i = \tfrac{1}{2}\sum_{\gamma=p+1}^{n}[E_\beta,E_\gamma]_i E_\gamma \in \Gamma\sigma Q.$$

Finally 2.12 is expressed by

$$(2.23)\qquad \alpha(E_\beta,E_\gamma) = 0 \quad \text{for} \quad p+1 \leq \beta,\ \gamma \leq n.$$

We return now to the general situation and define the tension field of a foliation F with respect to a Riemannian metric g_M. The trace $\mathrm{Tr}\,A(\nu)$ is linear in ν, hence $\mathrm{Tr}\,A \in \Gamma Q^*$. It follows that there exists a unique $\tau \in \Gamma Q$ such that

$$(2.24)\qquad \mathrm{Tr}\,A(\nu) = g_Q(\tau,\nu) \quad \text{for all} \quad \nu \in \Gamma Q.$$

We call τ the <u>tension field</u> of the foliation F.

From the matrix representation (2.16) we have clearly

$$(2.25)\qquad \mathrm{Tr}\,A(\nu) = \mathrm{Tr}\,W(\nu).$$

To evaluate τ at $x \in M$, let $e_1,\ldots,e_n \in T_xM$ be an orthonormal frame, such that $e_1,\ldots,e_p$ span L_x and $e_{p+1},\ldots,e_n$ span σQ_x. Then

$$(\mathrm{Tr}\,A(\nu))(x) = \sum_{i=1}^{p} g_M(A(\nu)e_i,e_i) = \sum_{i=1}^{p} g_Q(\alpha(e_i,e_i),\nu).$$

Comparing with (2.24), we obtain the evaluation formula

$$(2.26)\qquad \tau_x = \sum_{i=1}^{p}\alpha(e_i,e_i) \in Q_x.$$

In terms of a local orthonormal frame $E_1,\ldots,E_n$ of TM with $E_i \in \Gamma L$ $(i = 1,\ldots p)$ and $E_\beta \in \Gamma\sigma Q$ $(\beta = p+1,\ldots,n)$ we obtain by 2.19, part (i)

$$(2.27)\qquad \tau_x = \sum_{\beta=p+1}^{n}\ \sum_{i=1}^{p}[E_\beta,E_i]_i E_\beta\ .$$

Since α restricts on any leaf $L \subset M$ to its second fundamental form, it follows from (2.26) that $\frac{1}{p}\tau$ restricts on any leaf $L \subset M$ to its mean curvature vectorfield. As a consequence of the preceding discussion, we have the following result.

2.28 THEOREM. <u>Let</u> F <u>be a foliation on a manifold and</u> g_M <u>a Riemannian metric. Then the following conditions are equivalent:</u>

(i) <u>all leaves of the foliation are minimal submanifolds,</u>

(ii) <u>the tension field</u> τ <u>vanishes,</u>

(iii) $\mathrm{Tr}\, A(\nu) = 0$ <u>for all</u> $\nu \in \Gamma Q$,

(iv) $\mathrm{Tr}\, W(\nu) = 0$ <u>for all</u> $\nu \in \Gamma Q$.

The equivalence of (i) and (iv) is used by Rummler [RU1] in his study of geometrically taut foliations. Many characterizations of geometric tautness are known [HA], [S]. For compact foliations (all leaves are compact) Rummler [RU1], [RU2] shows the equivalence of geometric tautness with local stability of the foliation, i.e. any of the conditions in Epstein's equivalence theorem [EP1], [EP2].

It is of interest to comment on Rummler's characterization of the minimality by properties of the characteristic form χ_F [RU1]. He introduces this form $\chi_F \in \Omega^p(M)$ for a p-dimensional foliation F with oriented bundle L on a Riemannian manifold by

$$(2.29)\quad \begin{cases} \chi_F(X_1,\ldots,X_p) = 1 \text{ if } X_1,\ldots,X_p \text{ is an oriented orthonormal frame o} \\ \chi_F(Y_1,\ldots,Y_p) = 0 \text{ if one of the } Y_i \text{ is orthogonal to } L. \end{cases}$$

This form is well-defined and its differential satisfies the formula [RU1, p. 228]

$$(2.30)\qquad d\chi_F(X_1,\ldots,X_p;\, Y_\nu) = (-1)^{p+1}\mathrm{Tr}\, W(\nu)\cdot\chi_F(X_1,\ldots,X_p)$$

for $X_1,\ldots,X_p \in \Gamma L$ and $\nu \in \Gamma Q$, $Y_\nu = \sigma(\nu) \in \Gamma\sigma Q$.

The form χ_F depends by definition only on the metric $g_M|L$ and the orthogonal splitting σ. Thus if we start with a Riemannian foliation, if necessary we can replace $g_M|\sigma Q$ by a holonomy invariant metr. In particular, we get the following fact.

2.31 COROLLARY. <u>If a</u> R-<u>foliation is geometrically taut, there exist a bundle-like metric for which the leaves are minimal</u>.

One characterization of geometric tautness can be given in terms of the spectral sequence associated to a foliation [KT2]. For this purpose we consider the filtration of $\Omega^\bullet(M)$ defined by a foliation F on M, i.e.

$$(2.32)\quad \begin{cases} \omega \in F^r\Omega^m(M) \quad \text{if and only if} \\ \omega(X_1,\ldots,X_m) = 0 \quad \text{when } m-r+1 \text{ of the } X_i\text{'s are in } \Gamma L. \end{cases}$$

It is a decreasing filtration of $\Omega^\bullet(M)$ by differential ideals, which are closed under the operators $i(X)$ and $\theta(X)$ for $X \in \Gamma L$. Clearly $F^0\Omega^m = \Omega^m$, $F^{m+1}\Omega^m = 0$. For $p = \dim L$ further

(2.33) $$F^r\Omega^{p+r} = \Omega^{p+r},$$

since $\omega \in \Omega^{p+r}$ vanishes when $p + r - r + 1 = p + 1$ of the X_i's are in ΓL.

From the local decomposition of $\omega \in F^r\Omega^m$, it follows that

$$F^r\Omega^m = \Gamma(M,(\Lambda^r\mathcal{Q}^*\cdot\Omega_M)^m), \quad r \leq m$$

where Q^* denotes the dual normal bundle, $\mathcal{Q}^*$ its sheaf of sections, Ω_M the De Rham sheaf on M and $(\Lambda^r\mathcal{Q}^*\cdot\Omega_M)^m$ the forms of degree m in the subsheaf of Ω_M generated by $\Lambda^r\mathcal{Q}^*$ (see [KT2, section 5] for more details). It follows that

(2.34) $$F^{q+1}\Omega^m = 0, \quad q = \operatorname{codim} \mathcal{F} .$$

The $\mathcal{F}$-trivial forms $\omega \in \Omega^{p+r}(M)$ $(r\geq 0)$ [RU1] are now characterized by $\omega \in F^{r+1}\Omega^{p+r}(M)$ and the $\mathcal{F}$-closed forms by $d\omega \in F^{r+2}\Omega^{p+r+1}(M)$. Rummler's formula (2.30) implies now that the foliation $\mathcal{F}$ satisfies the equivalent conditions of Theorem 2.28 (minimality) if and only if $d\chi_{\mathcal{F}} \in F^2\Omega^{p+1}(M)$, i.e. $\chi_{\mathcal{F}}$ is $\mathcal{F}$-closed.

For the associated graded object we have $G^r\Omega^{r+s} = 0$ for $r > q$, $s > p$, and for $0 \leq r \leq q$, $0 \leq s \leq p$ we find that

$$E_0^{r,s} = G^r\Omega^{r+s} \cong \operatorname{Hom}(\Lambda^s L,\Lambda^r Q^*).$$

We use the notation

(2.35) $$T_L^s(\Lambda^r Q^*) = \operatorname{Hom}(\Lambda^s L,\Lambda^r Q^*)$$

to point out the resolvent character of the corresponding sheaf complex $T_L(\Lambda^r\mathcal{Q}^*)$ resolving $(\Lambda^r\mathcal{Q}^*)^L$, the basic r-forms of the foliation $\mathcal{F}$ (see KT2, section 5 , based on the general theory in [KT1]). The differential d_o corresponds under this identification to the Chevalley-Eilenberg differential d_C in $T_L(\Lambda^r Q^*)$, defined with respect to the ΓL-action on $\Gamma\Lambda^r Q^*$.

It follows that the E_1-term of the spectral sequence associated to the above filtration is given by [KT1], [KT2]

$$E_1^{r,s} \cong H^s(M,L;\Lambda^r\mathcal{Q}^*) = H^s(M,\Lambda^r\mathcal{Q}^{*L}).$$

For $s = p = \dim L$, there are canonical surjective maps

$$j : \Omega^{p+r}(M) \xrightarrow{j_1} E_o^{r,p} \xrightarrow{j_2} E_1^{r,p}, \quad r = 1,\dots,q.$$

In terms of this spectral sequence, geometric tautness can be characterized as follows (compare also [HA]).

2.36 PROPOSITION. A foliation given by $L \subset TM$ is geometrically tau if and only if there exists a volume form $\omega_o \in E_o^{o,p} = \mathrm{Hom}(\Lambda^p L, R)$ suc that $d_1 j_2(\omega_o) = 0$.

3. HARMONICITY

Let $E \to M$ be a vectorbundle over a Riemannian manifold, and ∇ a connection on E. The star operator on M extends to E-valued for

$$* : \Omega^r(M,E) \to \Omega^{n-r}(M,E), \qquad n = \dim M.$$

The codifferential $d_\nabla^* : \Omega^r(M,E) \to \Omega^{r-1}(M,E)$, $r > 0$ of the exterior differential d_∇ is given in terms of the star operator by

$$d_\nabla^*\omega = (-1)^{n(r+1)+1} *d_\nabla*\omega, \qquad \omega \in \Omega^r(M,E).$$

The evaluation formula for an orthonormal basis $e_1,\ldots,e_n \in T_xM$ is as follows:

$$(3.1) \qquad (d_\nabla^*\omega)_x(X_1,\ldots,X_{r-1}) = -\sum_{i=1}^n (\nabla_{e_i}\omega)_x(e_i;X_1,\ldots,X_{r-1})$$

where $X_1,\ldots,X_{r-1} \in T_xM$, and $(\nabla_{e_i}\omega)_x$ denotes the value at x of $\nabla_X\omega$ for any vectorfield X such that $X_x = e_i$, $i = 1,\ldots,r-1$.

The usual assumption for a satisfactory theory is a fiber metric g_E and a metric connection ∇ on E. As a consequence the codifferential d_∇^* becomes the formal adjoint of d_∇ with respect to the naturally induced scalar product on E-valued forms over a compact oriented manifold M. The kernel of the Laplacian

$$\Delta = d_\nabla d_\nabla^* + d_\nabla^* d_\nabla$$

coincides then precisely with the forms which are both d_∇-closed and d_∇^*-closed (see also the beginning of section 4 for comments on this point).

For the normal bundle Q of a foliation on M the connection ∇ on Q defined by a Riemannian metric g_M via (1.3) need not be metr with respect to g_Q, as we have seen. Thus we say that $\omega \in \Omega^r(M,Q)$ is harmonic if $d_\nabla\omega = 0$ and $d_\nabla^*\omega = 0$. In case ∇ is metric (and M compact and oriented), this condition is equivalent to $\Delta\omega = 0$ by the above adjointness argument (see e.g. [ES]). This will in particular be the case for R-foliations and bundle-like metrics g_M.

Denote as before by $\pi \in \Omega^1(M,Q)$ the canonical projection $TM \to Q$. Then we have the following fact.

3.2 PROPOSITION. <u>The tension field</u> τ <u>of</u> 2.24 <u>is given by</u>

$$\tau = d_\nabla^* \pi \ .$$

Proof. Consider an orthonormal basis $e_1,\dots,e_n \in T_xM$ with $e_1,\dots,e_p$ spanning L_x and $e_{p+1},\dots,e_n$ spanning σQ_x. By (3.1), (2.3) and (2.23) we have then

$$(d_\nabla^*\pi)_x = -\sum_{i=1}^{n} (\nabla_{e_i}\pi)_x(e_i) = \sum_{i=1}^{n} \alpha(e_i,e_i) = \sum_{i=1}^{p} \alpha(e_i,e_i).$$

But this is precisely the evaluation formula (2.26) for the tension field τ at x. ∎

3.3 THEOREM. <u>Let</u> F <u>be a foliation on a manifold and</u> g_M <u>a Riemannian metric. Then the following conditions are equivalent:</u>

(i) π <u>is a harmonic form,</u>

(ii) <u>all leaves of the foliation are minimal submanifolds.</u>

<u>If</u> F <u>is an</u> R-<u>foliation,</u> g_M <u>a bundle-like metric, and</u> M <u>compact and oriented, then these conditions are equivalent to</u>

(iii) $\Delta\pi = 0$.

Proof. $d_\nabla \pi = 0$ in any case by (2.2). The equivalence of (i) and (ii) follows from 3.2. Clearly (i) ⇒ (iii) without any assumption. To prove the converse under the stated conditions, we observe that ∇ on Q is then by Theorem 1.11, part (i) a metric connection with respect to the induced metric g_Q. Thus d_∇^* becomes the formal adjoint of d_∇ with respect to the canonical scalar product on Q-valued forms. From this it follows that a form satisfying $\Delta\omega = 0$ is necessarily d_∇-closed and d_∇^*-closed. ∎

A foliation is <u>harmonic</u> if it satisfies one (and hence all) of the equivalent conditions of Theorems 2.28 and 3.3.

An apparent weakening of the condition of the vanishing tension field $\tau \in \Gamma Q$ would be to require $\nabla\tau = 0$. This amounts to the requirement that the mean curvature vectorfield $1/p\cdot\tau$ of the leaves of the foliation is a ∇-parallel section of $\sigma Q \subset TM$. But in fact the following holds, at least for R-foliations.

3.4 PROPOSITION. <u>If</u> F <u>is a</u> R-<u>foliation, and</u> M <u>compact oriented, then</u>

$$\nabla\tau = 0 \Rightarrow \tau = 0.$$

Proof. Note that for the 0-form $\tau \in \Omega^0(M,Q)$ by definition $d_\nabla \tau = \nabla$
so that

$$\Delta\pi = d_\nabla d_\nabla^* \pi = d_\nabla \tau = \nabla\tau .$$

Thus

$$\nabla\tau = 0 \Rightarrow \Delta\pi = 0 \Rightarrow d_\nabla^* \pi = 0, \qquad \text{i.e.} \qquad \tau = 0. \qquad \blacksquare$$

The assumption in (3.4) is that the mean curvature of the leaves is parallel on all of M, not only along the leaves of the foliation. The latter (weaker) property would say that all leaves have $\mathring{\nabla}$-paralle mean curvature in M.

In this context it is of interest to look at the dual of the complex (2.35) for $r = 1$, namely

$$T_L^{\bullet}(Q) = \mathrm{Hom}(\Lambda^{\bullet} L, Q). \tag{3.5}$$

This is the restriction to L of the complex $\Omega^{\bullet}(M,Q)$. The Bott conn tion $\mathring{\nabla}$ in Q along L defines a differential $d_L : T_L^s(Q) \to T_L^{s+1}(Q)$ $s \geq 0$. The condition that $\mathring{\nabla}\tau = 0$ for the tension τ holds along th leaves of the foliation is then expressed by $d_L \tau = 0 \in T_L^2(Q)$.

Returning to the complex $\Omega^{\bullet}(M,Q)$, we observe that for $\omega \in \Omega^r(M,Q)$ we have $d_\nabla^2 \omega = R_\nabla \wedge \omega$, with $R_\nabla \in \Omega^2(M, \mathrm{End}\, Q)$ the curv ture of ∇. This curvature is zero when restricted to each leaf, but not on M. For a Riemannian foliation on a compact oriented manifold M there exists a decomposition of $\Omega^r(M,Q)$ [ES, p. 121]

$$\Omega^r(M,Q) = H^r(M,Q) \oplus \{ d_\nabla \Omega^{r-1}(M,Q) + d_\nabla^* \Omega^{r+1}(M,Q) \} , \tag{3.6}$$

where the space $H^r(M,Q)$ of harmonic forms is finite-dimensional. The first sum is orthogonal, but the sum in the parenthesis is not necessarily orthogonal.

In the remainder of this section we discuss examples of harmonic foliations. The simplest example is the foliation of $\mathbb{R}^n$ by p-pla parallel to a given p-dimensional subspace. A foliation of a Kähler manifold by complex submanifolds is harmonic. This follows from the fact that a complex submanifold of a Kähler manifold is necessarily minimal [KN, Vol. II, p. 380]. As an example let $P_1, \ldots, P_q$ be complex polynomials in $z = (z_1, \ldots, z_n) \in \mathbb{C}^n$. Let for $c = (c_1, \ldots, c_q)$

$$V_c = \{ z \mid P_i(z) = c_i \}.$$

If the rank of the matrix $\left(\frac{\partial P_i}{\partial z_j}\right)$ is q, then $V_c \subset \mathbb{C}^n$ is a complex submanifold of codimension q. Since the rank condition is an open condition, it follows that a neighborhood of V_c is harmonically foliated. It is interesting to observe that in this case not only $\mathrm{Tr}\, A(\nu) = 0$, but in fact all odd elementary symmetric functions of the eigenvalues vanish [L].

3.7 FOLIATIONS OF CODIMENSION ONE. Let F be of codimension 1 and assume F to be transversely orientably by a nowhere zero section $s \in \Gamma Q$. Let g_M be a Riemannian metric on M. Through renormalization we can assume that s is of unit length for the induced metric g_Q in Q. Let $Z = \sigma(s) \in \Gamma\sigma Q$. Then $g_M(Z,Z) = 1$. An associated form $\omega \in \Omega^1(M)$ is defined by

$$\omega(X) = g_M(X,Z) \quad \text{for} \quad X \in \Gamma TM.$$

Clearly ω is non-singular and $L = \ker \omega$. Since $\omega(Z) = g_M(Z,Z) = 1$, it follows that ω is the 1-form dual to the vectorfield Z. If η_M is the volume form associated to g_M, it follows that

$$(3.8) \qquad *\omega = i(Z)\eta_M \equiv \chi_F \in \Omega^{n-1}(M)$$

is Rummler's characteristic form.

In the following we identify σQ and Q. Since the projection $\pi : TM \to Q$ is orthogonal, we have $\pi(X) = \omega(X)\cdot Z$. The non-zero section $Z \in \Gamma Q$ identifies Q with the trivial bundle $M \times \mathbb{R} \to M$ and $\Omega^r(M,Q)$ with $\Omega^r(M)$. But the connection ∇ in Q need not correspond to the trivial connection in $M \times \mathbb{R} \to M$. With these notations we have then the following characterizations of the harmonicity of F.

3.9 PROPOSITION. Let F be a transversally oriented foliation of codimension 1. Then the following conditions are equivalent:

(a) F is harmonic;

(b) $d^*_\nabla \pi = 0$;

(c) $d^*\omega = 0$;

(d) $\mathrm{Tr}\, W(Z) = 0$;

(e) $d\chi_F = 0$ (Rummler's criterion);

(f) $\mathrm{div}\, Z = 0$.

Proof. The equivalence of (a), (b) and (d) has been established for any codimension. (e) $\Rightarrow$ (d) by formula (2.30). For codimension 1 the same formula shows conversely that (d) $\Rightarrow$ (e). The equivalence (e) $\Leftrightarrow$ (f) follows from (3.8) and

$$d\chi_F = di(Z)\eta_M = \theta(Z)\eta_M = \mathrm{div}(Z)\cdot\eta_M .$$

It suffices to establish (b) ⇔ (c). Let $E_1,\dots,E_{n-1},Z$ be a local fram of $TM \cong L \oplus \sigma Q$ extending an orthonormal basis $e_1,\dots,e_n \in T_xM$ with $e_1,\dots,e_{n-1} \in L_x$ and $e_n = Z_x \in Q_x$. By (3.1), (2.3)

$$\begin{aligned}(d_\nabla^*\pi)_x &= -\sum_{i=1}^{n} (\nabla_{e_i}\pi)_x(e_i)\\ &= -\sum_{i=1}^{n} (\nabla_{e_i}(\pi(E_i)) - \pi(\nabla^M_{e_i}E_i)).\end{aligned}$$

But $\pi(E_i) = 0$ for $1 \leq i \leq n-1$. For $i = n$ the term on the RHS is the value at x of the vectorfield $\pi(\nabla^M_Z Z) - \nabla_Z Z$, which is 0 by (1.3). Thus

$$(d_\nabla^*\pi)_x = \sum_{i=1}^{n-1} \pi(\nabla^M_{e_i}E_i). \tag{3.10}$$

On the other hand, we have for the ordinary codifferential

$$\begin{aligned}(d^*\omega)_x &= -\sum_{i=1}^{n} (\nabla_{e_i}\omega)(e_i)\\ &= -\sum_{i=1}^{n} (e_i.\omega(E_i) - \omega(\nabla^M_{e_i}E_i)).\end{aligned}$$

Here $\omega(E_i) = 0$ for $i = 1,\dots,n-1$. For $i = n$ we have $\omega(Z) = 1$, so that $Z\omega(Z) = 0$. Further by definition of ω

$$\omega(\nabla^M_Z Z) = g_M(\nabla^M_Z Z, Z).$$

But $0 = Zg_M(Z,Z) = 2g_M(\nabla^M_Z Z,Z)$. Thus $\omega(\nabla^M_Z Z) = 0$. It follows that

$$(d^*\omega)_x = \sum_{i=1}^{n-1} \omega(\nabla^M_{e_i}E_i). \tag{3.11}$$

Since $\pi(X) = \omega(X).Z$, (3.10) and (3.11) show the equivalence (b) ⇔ (c)

From the proof we further see that

$$0 = \omega(\nabla^M_Z Z) = g_M(\nabla^M_Z Z, Z) = g_Q(\nabla_Z Z, Z)$$

which implies

$$\nabla_Z Z = 0. \tag{3.12}$$

3.13 COROLLARY. <u>Let F be a transversely oriented foliation of codimension 1. Then the following conditions are equivalent:</u>

(i) F <u>is geometrically taut;</u>

(ii) <u>there exists a volume form η_M and a divergence free transversal vectorfield Z on M.</u>

Proof. (i) ⇒ (ii) follows from Proposition 3.9. (ii) ⇒ (i): It suffices to find a Riemannian metric g_M such that η_M is its volume form, Z of unit length and L orthogonal to the line bundle Q spanned by Z. But $TM = L \oplus Q$ and $\theta(Z)\eta_M = \mathrm{div}(Z)\cdot\eta_M = 0$ implies that η_M belongs to a $SL(n-1)$-reduction of the $GL(n-1) \times GL(1)$-structure on TM. Reducing to $SO(n-1)$ yields g_M as desired. ∎

In the discussion above the foliation F was not assumed to be Riemannian. The role of this assumption is clarified by the following result. We would like to acknowledge a helpful conversation with T. Duchamp on this point.

3.14 PROPOSITION. Let F be a transversally oriented foliation of codimension 1 and g_M a Riemannian metric on M. Define g_Q, Z and ω as above. Then the following conditions are equivalent;

(a) $\overset{\circ}{\nabla}_X g_Q = 0$ for all $X \in \Gamma L$, i.e. F is Riemannian;

(b) $d\omega = 0$;

(c) $\theta(X)\omega = 0$ for all $X \in \Gamma L$;

(d) $\overset{\circ}{\nabla}{}^*_X\omega = 0$ for all $X \in \Gamma L$;

(e) $\overset{\circ}{\nabla}_X Z = 0$ for all $X \in \Gamma L$;

(f) $\nabla Z = 0$.

Proof. For $X \in \Gamma L$ we have by [KT3, p. 26]

$$\overset{\circ}{\nabla}{}^*_X\omega = \theta(X)\omega = i(X)d\omega. \tag{3.15}$$

This 1-form vanishes on L. Namely for $Y \in \Gamma L$

$$i(Y)\overset{\circ}{\nabla}{}^*_X\omega = i(Y)\theta(X)\omega = i(Y)i(X)d\omega = X\omega(Y) - Y\omega(X) - \omega[X,Y] = 0$$

Since $d\omega(Z,Z) = 0$, it follows

$$d\omega = 0 \Leftrightarrow i(Z)i(X)d\omega = 0 \quad \text{for} \quad X \in \Gamma L. \tag{3.16}$$

This establishes (b) ⇔ (c) ⇔ (d) by (3.15).

Next we observe that

$$i(Z)\overset{\circ}{\nabla}{}^*_X\omega = i(Z)\theta(X)\omega = i(Z)i(X)d\omega = X\omega(Z) - Z\omega(X) - \omega[X,Z]. \tag{3.17}$$

The only nontrivial term is

$$-\omega[X,Z] = -\omega(\pi[X,Z]) = -\omega(\overset{\circ}{\nabla}_X Z) = -g_Q(\overset{\circ}{\nabla}_X Z,Z) = \tfrac{1}{2}(\overset{\circ}{\nabla}_X g_Q)(Z,Z).$$

Thus

$$d\omega(X,Z) = -g_Q(\overset{\circ}{\nabla}_X Z,Z) = \tfrac{1}{2}(\overset{\circ}{\nabla}_X g_Q)(Z,Z). \tag{3.18}$$

This establishes (b) ⇔ (e) ⇔ (a).

It remains to establish (e) ⇔ (f). Clearly (f) ⇒ (e). The converse follows from (3.12). ■

3.19 SUMMARY. Let F be a transversally oriented foliation of codimension 1 and g_M a Riemannian metric on M. Let $\omega(X) = g_M(X,Z$ Then the following characterizations hold:

(i) F is harmonic iff $d^*\omega = 0$;

(ii) F is Riemannian iff $d\omega = 0$;

(iii) F is harmonic Riemannian iff ω is harmonic.

A simply connected compact manifold cannot support a Riemannian foliation of codimension one, since a closed 1-form has necessarily zeros. In 3.34 below we give an example of a harmonic foliation of codimension 1 which is not Riemannian.

3.20 PRINCIPAL FIBRATIONS. Let $G \to P \xrightarrow{\pi} B$ be a principal G-bundl with connection form ω. A metric g on the Lie algebra $\mathfrak{g}$ and a metric g_B on B define a bundle-like metric g_P on P. Let $X,Y \in \Gamma TB$ and $\tilde{X},\tilde{Y}$ the horizontal lifts with respect to ω. Let further ξ^*, η^* denote the fundamental vectorfields on P correspond to $\xi,\eta \in \mathfrak{g}$. Then g_P is characterized by the formulas

$$g_P(\tilde{X},\tilde{Y}) = g_B(X,Y) \qquad \text{for} \qquad X,Y \in \Gamma TB,$$

$$g_P(\xi^*,\eta^*) = g(\xi,\eta) \qquad \text{for} \qquad \xi,\eta \in \mathfrak{g},$$

$$g_P(\xi^*,\tilde{X}) = 0.$$

This turns P into a Riemannian foliation with bundle-like metric g_P Its normal bundle is the horizontal distribution $H = \ker \omega$.

3.21 PROPOSITION. The second fundamental form α defined by (2.3) i characterized by the following formulas:

(i) $\alpha(\tilde{X},\tilde{Y}) = 0$ for $X,Y \in \Gamma TB$;

(ii) $\alpha(\xi^*,\eta^*) = 0$ for $\xi,\eta \in \mathfrak{g}$;

(iii) $g_P(\alpha(\xi^*,\tilde{X}),\tilde{\nu}) = \frac{1}{2}\, g(\Omega(\tilde{X},\tilde{\nu}),\xi)$ for $\xi \in \mathfrak{g}$, X and $\nu \in \Gamma T$ and Ω the curvature form of ω on P.

Proof. (i) follows from 2.12. To establish (ii), we use 2.18 and fi for $\nu \in \Gamma TB$ by linear expansion

$$g_P(\alpha(\xi^*,\eta^*),\tilde{\nu}) = -\tfrac{1}{2}(g_P([\xi^*,\nu],\eta^*) + g_P([\eta^*,\tilde{\nu}],\xi^*))$$

$$= -\tfrac{1}{2}(g(\omega[\xi^*,\tilde{\nu}],\eta) + g(\omega[\eta^*,\tilde{\nu}],\xi)).$$

For the curvature Ω we have $i(\xi^*)\Omega = 0$. By the structure equation on the other hand

$$\Omega(\xi^*,\tilde{\nu}) = \xi^*\omega(\tilde{\nu}) - \tilde{\nu}\omega(\xi^*) - \omega[\xi^*,\tilde{\nu}] + [\omega(\xi^*),\omega(\tilde{\nu})] = -\omega[\xi^*,\tilde{\nu}].$$

Thus $\omega[\xi^*,\tilde{\nu}] = 0$ and similarly $\omega[\eta^*,\tilde{\nu}] = 0$. It follows that $\alpha(\xi^*,\eta^*) = 0$, i.e. $A_1 = 0$ in the terminology of 2.11. To verify (iii), we obtain from 2.13

$$g_P(\alpha(\tilde{X},\xi^*),\tilde{\nu}) = \tfrac{1}{2}\, g_P([\tilde{\nu},\tilde{X}],\xi^*) = \tfrac{1}{2}\, g(\omega[\tilde{\nu},\tilde{X}],\xi),$$

and from the structure equation

$$\Omega(\tilde{X},\tilde{\nu}) = \tilde{X}\omega(\tilde{\nu}) - \tilde{\nu}\omega(\tilde{X}) - \omega[\tilde{X},\tilde{\nu}] + [\omega(\tilde{X}),\omega(\tilde{\nu})] = -\omega[\tilde{X},\tilde{\nu}].$$

Thus

$$g_P(\alpha(\xi^*,\tilde{X}),\tilde{\nu}) = \tfrac{1}{2}\, g(\Omega(\tilde{X},\tilde{\nu}),\xi). \quad ■$$

3.22 COROLLARY. <u>The operator</u> $A(\tilde{\nu})$ <u>defined by</u> (2.8) <u>for</u> $\nu \in \Gamma TB$ <u>has the matrix representation</u>

$$A(\tilde{\nu}) = \begin{pmatrix} 0 & A_2(\tilde{\nu}) \\ A_2^*(\tilde{\nu}) & 0 \end{pmatrix}$$

with $A_2(\tilde{\nu})\tilde{X} = \tfrac{1}{2}\,\Omega(\tilde{X},\tilde{\nu})^* = \tfrac{1}{2}\,(\omega[\tilde{\nu},\tilde{X}])^*$ for $X \in \Gamma TB$.

It follows that the fibers are totally geodesic submanifolds, a well known result. Thus a principal fibration defines a harmonic foliation. It was already established in [ES] that the projection $\pi : P \to B$ is a harmonic map. Note that by the expression for A_2 given in 3.22 we have $A_2 = 0$ iff ω is a flat connection on P. Thus, e.g., the coset foliation of G by a closed subgroup H is always harmonic for an appropriate metric. If we assume that the Lie algebra $\mathfrak{h} \subset \mathfrak{g}$ has an H-invariant complement m ($\mathfrak{g} = \mathfrak{h} \oplus m$), we have $A_2 = 0$ iff m is an ideal, i.e. the projection $\theta : \mathfrak{g} \to \mathfrak{h}$ a Lie homomorphism, i.e. $\mathfrak{g}$ a semi-direct product $\mathfrak{h} \times_\varphi m$.

If $\Gamma \subset \mathrm{Aut}(P)$ is a discrete group of isometric bundle-automorphisms, acting freely on P (but not necessarily on B), the fibration $P \xrightarrow{\pi} B$ induces a natural foliation F of codimension $= \dim B$ on the orbit-space $\Gamma\backslash P$. For the induced metric on $\Gamma\backslash P$, the foliation F is still totally geodesic. In this way, one obtains examples of minimal or harmonic foliations not given by fibrations or submersions. In the case of a coset-fibration $G \xrightarrow{\pi} G/H$ one has to assume the existence of an H-invariant metric g on the Lie-algebra $\mathfrak{g}$.

3.23 QUOTIENT FOLIATIONS. Let $K \to P \xrightarrow{\pi} B$ be a principal K-bundle with compact group K, and ω a connection form on P. Let $L \subset TB$ be a codimension q foliation on B with normal bundle Q. Assume L to be a Riemannian foliation and g_B a bundle-like metric on B. Let $\tilde{L} \subset TP$ be the pull-back foliation with normal bundle $\tilde{Q} \cong \pi^* Q$. Let α^K, A^K denote the second fundamental form and associated operato[r] for the foliation $K \to P \xrightarrow{\pi} B$, α and A the corresponding objects for the foliation L on B, $\tilde{\alpha}$ and $\tilde{A}$ the corresponding objects fo[r] the foliation $\tilde{L}$ on P. The operator $A(\nu) : TB \to TB$ for $\nu \in \Gamma Q$ lifts to $(\pi^* A)(\pi^* \nu) : \pi^* TB \to \pi^* TB$ by

$$(\pi^* A)(\pi^* \nu)\tilde{X} = \pi^*(A(\nu)X) \qquad \text{for} \qquad X \in \Gamma TB.$$

The following result is then easily established.

3.24 PROPOSITION. $\tilde{\alpha}$ <u>is characterized by</u> α <u>via the following</u> <u>formulas</u>:

    (i) $\tilde{\alpha}(\tilde{X},\tilde{Y}) = \pi^* \alpha(X,Y)$ <u>for</u> $X,Y \in \Gamma TB$;

    (ii) $\tilde{\alpha}(\xi^*, \eta^*) = 0$ <u>for</u> $\xi, \eta \in \mathcal{G}$;

    (iii) $g_p(\tilde{\alpha}(\xi^*,\tilde{X}),\pi^* \nu) = \frac{1}{2} g(\Omega(\tilde{X},\pi^* \nu),\xi)$ for $\xi \in \mathcal{G}$, $X \in \Gamma TB$,

$\nu \in \Gamma TQ$, <u>and</u> Ω <u>the curvature of</u> ω.

   $\tilde{A}$ <u>is characterized by</u> $A^K(\nu)$ <u>and</u> A <u>via the matrix representa[tion]</u>

$$\tilde{A}(\pi*\nu) = \left(\begin{array}{c|c} 0 & A_2^K(\pi^* \nu) \\ \hline (A_2^K(\pi^* \nu))^* & (\pi^* A)(\pi^* \nu) \end{array} \right)$$

<u>with respect to the decomposition</u> $TP \cong (P \times k) \oplus \pi^* TB$, <u>where</u>

$$(\pi^* A)(\pi^* \nu) = \left(\begin{array}{c|c} \pi^* A_1(\nu) & \pi^* A_2(\nu) \\ \hline (\pi^* A_2(\nu))^* & 0 \end{array} \right)$$

<u>with respect to the decomposition</u> $\pi^* TB \cong \pi^* L \oplus \pi^* Q$.

   It follows that α, A for the foliation $L \subset TB$ are determine[d] by $\tilde{\alpha}$, $\tilde{A}$ for the foliation $\tilde{L} \subset TP$. In particular we get the follo[wing] consequence .

3.25 COROLLARY. <u>For the tension fields</u> $\tau \in \Gamma Q$ <u>of</u> $L \subset TB$ <u>and</u> $\tilde{\tau} \in \Gamma\tilde{Q}$ <u>of</u> $\tilde{L} \subset TP$ <u>we have</u> $\tilde{\tau} = \pi^* \tau$. <u>Thus</u> $L \subset TB$ <u>and</u> $\tilde{L} \subset TP$ <u>ar</u>[e] <u>simultaneously harmonic foliations.</u>

3.26 FIBRATIONS WITH HOMOGENOUS FIBERS. Let $G \to P \xrightarrow{\pi} X$ be a principal G-bundle, $K \subset G$ a compact subgroup, and $M = P/K$. Then we have a commutative diagram

$$\begin{array}{ccc} & \overset{\pi'}{\nearrow} & P/K = M \\ P & & \downarrow \hat{\pi} \\ & \underset{\pi}{\searrow} & P/G = B \end{array}$$

and $\hat{\pi} : M \to B$ is a fibration with homogenous fiber G/K. The foliation $L \subset TM$ defined by $\hat{\pi}$ is the quotient foliation of $\tilde{L} \subset TP$ defined by π in the sense of 3.23. For the operator $\tilde{A}(\pi^*\nu)$ we have then by combining 3.22 and 3.24 the following matrix representation with respect to the decomposition $TP \cong (P \times k) \oplus T(\pi') \oplus \pi^*TB$:

$$\tilde{A}(\pi^*\nu) = \left(\begin{array}{c|c|c} 0 & 0 & A_2^K \\ \hline 0 & 0 & \pi'^*A_2 \\ \hline (A_2^K)^* & (\pi'^*A_2)^* & 0 \end{array} \right)$$

It follows from 2.6, 2.11 that the leaves of the quotient foliation $L \subset TM$ are totally geodesic submanifold, hence a fortiori minimal. Thus $L \subset TM$ is a harmonic foliation.

3.27 FIBRE SPACES. Let $F \to E \xrightarrow{\pi} B$ be a smooth locally trivial fiber space with connected base and fiber spaces B and F. We can view it as a foliation F with tangent bundle sequence

$$0 \to T(\pi) \to TE \underset{\pi_*}{\overset{\sigma}{\rightleftarrows}} \pi^*TB \to 0 . \tag{3.28}$$

A Riemannian metric g_B on B determines a holonomy invariant metric $g_Q = \pi^*g_B$ on the normal bundle $Q = \pi^*TB$. With respect to a splitting σ of π_* in (3.28) g_Q extends to a bundle-like metric

$$g_E = g_{T(\pi)} \oplus g_Q \tag{3.29}$$

on $TE = T(\pi) \oplus \sigma Q$ for any choice of metric $g_{T(\pi)}$ on the tangent bundle $T(\pi)$ along the fiber of π. The foliation F is certainly locally stable, since B plays the role of the leaf space and is Hausdorff. By Rummler [RU2] F is then geometrically taut, at least when the fiber F is compact. By 2.31 there exists therefore a bundle like metric on E for which the fibers are minimal.

The preceding examples are all harmonic R-foliations. The following construction will provide a class of harmonic (even totally geodesic) foliations, which are not necessarily Riemannian.

3.30 FLAT FIBER SPACES. A flat structure on the fiber space $E \xrightarrow{\pi} B$ is given by a foliation $F^{\perp}$ transverse to the fibers of π, i.e. by a splitting σ, such that $\sigma Q \subset TE$ is involutive. It is equivalent to require that the horizontal lift $X \to \tilde{X} = \sigma(\pi^* X)$ of vectorfields X on B to projectable vectorfields $\tilde{X}$ on E preserves Lie bracke In the following we will assume that E is equipped with a bundle-li metric g_E defined as in (3.29) with respect to such a σ.

3.31 PROPOSITION. <u>For such a metric</u> g_E <u>on</u> E <u>the foliation</u> $F^{\perp}$ <u>transverse to the fibers of</u> $\pi : E \to B$ <u>is totally geodesic</u>.

Proof. By (2.15) it suffices to show that $\alpha^{\perp}(\tilde{X},\tilde{Y}) = 0$ for the second fundamental form $\alpha^{\perp}$ of $F^{\perp}$ and X, Y corresponding to $X,Y \in \Gamma TB$. By (2.3)

$$\alpha^{\perp}(\tilde{X},\tilde{Y}) = \pi_*^{\perp}(\nabla^E_{\tilde{X}}\tilde{Y}).$$

In terms of the second fundamental form α and the associated operat A of F, we have clearly

$$\pi_*^{\perp}(\nabla^E_{\tilde{X}}\tilde{Y}) = A_2(\tilde{Y})\tilde{X} .$$

But F is Riemannian, and thus by (2.13)

$$A_2(\tilde{Y})\tilde{X} = -\tfrac{1}{2}\, \pi_*^{\perp}[\tilde{X},\tilde{Y}] .$$

Thus

$$\alpha^{\perp}(\tilde{X},\tilde{Y}) = -\tfrac{1}{2}\pi^{\perp}[\tilde{X},\tilde{Y}] .$$

Since $\sigma Q \subset TE$ is involutive, this terms vanishes. By (2.15) the fo ation $F^{\perp}$ is therefore totally geodesic. ∎

Let B be orientable and η_B the volume form of g_B. Then $\eta_Q = \pi^*\eta_B$ is a volume form on $Q = \pi^* TB$. The form $\chi_{F^{\perp}} = \eta_Q$ is a characteristic form for $F^{\perp}$.

3.32 FLAT BUNDLES. A particular case of the preceding situation occurs when

$$E = \tilde{B} \times_{\Gamma} F \xrightarrow{\pi} B \tag{3.33}$$

where the fundamental group $\Gamma = \pi_1(B)$ acts on F via the holonomy homomorphism $h : \Gamma \to \mathrm{Diff}(F)$ of $F^{\perp}$. A (smooth) flat fiber space with compact fiber F is necessarily of this form. The normal bundl $Q^{\perp}$ of $F^{\perp}$ (the tangent bundle along the fiber of π) is then of th form

$$Q^{\perp} \cong \tilde{B} \times_{\Gamma} TF$$

with respect to the induced differential action $dh : \Gamma \to \text{Bundle Iso}(TF)$. From this it follows that the holonomy invariant metrics on $Q^{\perp}$ correspond to Riemannian metrics on F for which Γ acts by isometries. In order to construct a harmonic, but not Riemannian foliation, it suffices therefore to start with an action of Γ on F which is not metrizable for any Riemannian metric. Consider e.g. the canonical action of $G = SL(q+1,\mathbb{R})$ on the q-sphere $F = S^q$ realized by oriented rays in $\mathbb{R}^{q+1}$. Then $F \cong G/H$. For a cocompact discrete subgroup $\Gamma \subset G$ the canonical action $h : \Gamma \to \text{Diff}(S^q)$ is then not metrizable, and thus the totally geodesic foliation $F^{\perp}$ transverse to the fiber of

$$SL(q+1,\mathbb{R}) X_{\Gamma} S^q \to \Gamma \backslash SL(q+1,\mathbb{R})$$

not Riemannian.

3.34 PROPOSITION. Roussarie's foliàtion is harmonic, but not Riemannian.

Proof. This is the well-known foliation of the compact manifold $\Gamma\backslash SL(2,\mathbb{R})$, $\Gamma \subset SL(2,\mathbb{R})$ a discrete and cocompact subgroup, described in [GV]. It is harmonic for an appropriate metric on $\Gamma\backslash SL(2)$ by Proposition 3.31, since it is a foliation transverse to the fibers of a flat bundle. The flat structure on $\Gamma\backslash SL(2)$ is visible from the identifications

$$(3.35) \qquad \Gamma\backslash SL(2) \cong \Gamma\backslash SL(2) \times_{SO(2)} S^1 \cong SL(2)/SO(2) \times_{\Gamma} S^1 .$$

(This is a case of the flat bundle structure discussed on p. 181 of [KT3]. In the notation of that context, $\bar{G} = SL(2)$ and $\bar{K}/K = SO(2)/\{e\} \cong S^1$. We refer also to [KT3] for the Γ-actions involved on the various terms in (3.35).) This foliation has a non-trivial Godbillon-Vey class [GV]. By Proposition 3.14 it follows that it cannot be Riemannian. Namely condition (b) would then necessarily imply the triviality of the Godbillon-Vey class. ∎

3.36 HARMONIC FOLIATIONS AND CHARACTERISTIC CLASSES. The previous examples show that the characteristic classes of harmonic (or even totally geodesic) foliations need not be trivial.

The non-Riemannian foliations $F^{\perp}$ discussed in 3.30 provide examples of totally geodesic foliations with non-trivial characteristic homomorphism

$$\Delta_*(F^\perp) : H^\bullet(W(\mathfrak{gl}(q),O(q))_q) \to H^\bullet_{DR}(E), \qquad q = \dim F$$

(see [KT3, Ch. 7], [KT5, §5]). Furthermore, there are families of harmonic foliations with continuously variable characteristic classes [HI

In the case of R-foliations, the fibrations discussed in 3.20 and 3.26 provide examples of totally geodesic R-foliations with non-trivial characteristic homomorphism

$$\Delta_* : H(W(so(q))_{q'}) \to H^\bullet(M),$$

where $q = \dim B$ and $q' = q/2$ (see [KT3, Ch. 7], [KT4, §6] and [LP]). Of course, the characteristic classes coming from $W(\mathfrak{gl}(q))_q$ are necessarily zero in the Riemannian case.

4. ENERGY

Let $E \to M$ be a vectorbundle over a compact oriented manifold. A fiber metric g_E on E defines a bilinear map from $E \otimes E$ to the trivial line bundle on M. With respect to this map there is then a pairing

$$\Omega^r(M,E) \otimes \Omega^{r'}(M,E) \to \Omega^{r+r'}(M,R)$$

denoted $\omega \otimes \omega' \to g_E(\omega \wedge \omega')$. The scalar product on $\Omega^r(M,E)$ is then given by

$$\langle\omega,\omega'\rangle = \int_M g_E(\omega \wedge *\omega').$$

Let now ∇ be a connection on E. If ∇ is metric, i.e. satisfies $\nabla g_E = 0$, then the following derivation rule holds with respect to the pairing above:

$$dg_E(\omega \wedge \omega') = g_E(d_\nabla\omega \wedge \omega') + (-1)^r g_E(\omega \wedge d_\nabla\omega')$$

for $\omega \in \Omega^r(M,E)$ and $\omega' \in \Omega^{r'}(M,E)$. This rule in turn implies that the codifferential d_∇^* defined as the star-conjugate of d_∇ (see the beginning of section 3) is the formal adjoint of d_∇ with respect to the scalar product defined above.

We return to the general situation of the scalar product defined by a fiber metric g_E on $E \to M$. We will need the following discussion of densities. Let η denote the canonical volume form on M given by g_M. Then for $\omega,\omega' \in \Omega^r(M,E)$

$$g_E(\omega \wedge *\omega') = \bar{\rho}(\omega,\omega')\cdot\mu \in \Omega^n(M,R).$$

If we set $\rho(\omega) \equiv \bar{\rho}(\omega,\omega)$ for the density of ω, then

$$\|\omega\|^2 = \langle\omega,\omega\rangle = \int_M \rho(\omega)\mu \quad \text{for} \quad \omega \in \Omega^r(M,E).$$

The following description of the density $\rho(\omega)$ for $\omega \in \Omega^1(M,E)$ is useful. Define the endomorphism $B_\omega : TM \to TM$ by

(4.1) $$g_E(\omega(X),\omega(Y)) = g_M(B_\omega X,Y) \quad \text{for} \quad X,Y \in \Gamma TM.$$

Then the following holds.

4.2 PROPOSITION. $\rho(\omega) = \operatorname{Tr} B_\omega$.

Proof. We use $\Omega^1(M,E) \cong (T^*M \otimes E)$. Let $e_1,\ldots,e_n$ be an oriented orthonormal frame of T_xM with dual frame $e_1^*,\ldots,e_n^*$. The star operator satisfies

$$*e_i^* = (-1)^{i+1}e_1^* \wedge\ldots\wedge \hat{e}_i^* \wedge\ldots\wedge e_n^*, \text{ so that } e_i^* \wedge (*e_i^*) = e_1^* \wedge\ldots\wedge e_n^* \equiv \mu_x.$$

Further $*1 = \mu_x$, $*\mu = 1$.

Let $\omega = \alpha \otimes \beta \in \Gamma T^*M \otimes \Gamma E$. Then $\alpha_x = \sum_{i=1}^n \alpha_i e_i^*$ and

$$*\omega_x = \sum_{i=1}^n (-1)^{i+1}\alpha_i \cdot e_1^* \wedge\ldots\wedge \hat{e}_i^* \wedge\ldots\wedge e_n^* \otimes \beta.$$

Applying this to $\omega_x' = (\alpha' \otimes \beta')_x = (\sum_{i=1}^n \alpha_i' e_i^*) \otimes \beta_x'$, we get

$$(g_E(\omega \wedge *\omega'))_x = \sum_{j,i} (-1)^{i+1}\alpha_j\alpha_i' g_E(\beta,\beta')e_j \wedge (e_1^* \wedge\ldots\wedge \hat{e}_i^* \wedge\ldots\wedge e_n^*) = \sum_i \alpha_i\alpha_i' g_E(\beta,\beta')\mu.$$

This shows that for $\omega = \alpha \otimes \beta$, $\omega' = \alpha' \otimes \beta'$

$$\bar{\rho}(\omega,\omega') = g_M^*(\alpha,\alpha')\, g_E(\beta,\beta').$$

In particular

$$\rho(\omega) = g_M^*(\alpha,\alpha)\cdot g_E(\beta,\beta).$$

Let now $a_1,\ldots,a_q$ be an orthonormal frame of E_x. Then $\omega_x = \sum_{i,j} \alpha_{ij} e_i^* \otimes a_j$, and by the calculation above

$$\rho(\omega)_x = \sum_{i,j} \alpha_{ij}^2 \quad (i = 1,\ldots,n;\ j = 1,\ldots,q).$$

But on the other hand

$$(\mathrm{Tr}\ B\omega)_x = \sum_{k=1}^{n} g_M(B_\omega e_k, e_k) = \sum_{k=1}^{n} g_E(\omega(e_k),\omega(e_k)) = \sum_{k=1}^{n} g_E(\sum_{j=1}^{q} \alpha_{kj}\, a_j, \sum_{j=1}^{q} \alpha_{kj} a_j)$$

$$= \sum_{i,j} \alpha_{ij}^2 \qquad (i = 1,\ldots,n;\ j = 1,\ldots,q)$$

which proves the proposition. ■

We apply this to the normal bundle Q of a R-foliation F on a compact oriented manifold. For any Riemannian metric g_M on M and any fiber metric g_Q on Q, we define the <u>energy of the foliation</u> by

$$(4.3) \qquad E(F) = \tfrac{1}{2} \|\pi\|^2,$$

where $\pi \in \Omega^1(M,Q)$ is the canonical projection $\pi : TM \to Q$. For the foliation defined by a Riemannian submersion $f : M \to N^q$, this is the energy of f as defined by Eells-Sampson [ES].

In this definition g_M and g_Q are not necessarily related. Consider now the orthogonal decomposition $TM \cong L \oplus L^\perp$ with respect to g_M, and the corresponding splitting $\sigma : Q \to L^\perp \subset TM$ of (1.1).

4.4 PROPOSITION. <u>Assume that</u> g_Q <u>on</u> Q <u>is induced by</u> g_M. <u>Then</u>

$$E(F) = \tfrac{1}{2}\cdot q\cdot \mathrm{Vol}(M).$$

Proof. Let $e_1,\ldots,e_n$ be an orthonormal frame of T_xM such that $e_1,\ldots,e_p$ span L_x and $e_{p+1},\ldots,e_n$ span Q_x. Let $e_1^*,\ldots,e_n^*$ be the dual frame of T_x^*M. Then

$$\pi_x = \sum_{\gamma=p+1}^{n} e_\gamma^* \otimes e_\gamma \in T_x^*M \otimes \sigma Q_x.$$

Clearly the operator $B_\pi : TM \to TM$ in (4.1) is the projection operator onto σQ. By (4.2) it follows that $\rho(\pi) = q$. Thus $\|\pi\|^2 = \int_M \rho(\pi)\mu = q \int_M$ as claimed. ■

Note that this result holds in particular for a bundle-like metric g_M on TM and the induced metric g_Q on Q. If the R-foliation F undergoes a variation through R-foliations F_t, $t \in (-\varepsilon,\varepsilon)$, $F_0 = F$, the fixed metric g_M will generally fail to be bundle-like for F_t, $t \neq 0$. Thus the energy $E(F_t)$ will generally be a non-constant function. We wish to show that for a certain type of variation a R-foliation F is an extremal of the energy functional precisely when π is harmonic.

To explain this, we need the description of a R-foliation by a Haefliger cocycle $\{U_\alpha, f^\alpha, \gamma^{\alpha\beta}\}$. The U_α cover M, the $f^\alpha : U_\alpha \to V_\alpha \subset N$ are submersions onto open submanifolds of a Riemanniar manifold N, and on $U_{\alpha\beta} = U_\alpha \cap B_\beta$

(4.5) $$f^\alpha = \gamma^{\alpha\beta} \circ f^\beta$$

with local isometries $\gamma^{\alpha\beta}$ of N. On U_α, $Q = (f^\alpha)^*TN$ and $g_Q = (f^\alpha)^* g_N$.

Let M be compact. We consider a refinement $\mathcal{U}'$ of the covering $\mathcal{U}$ with relative compact U'_α. A variation $\mathcal{F}_t$ of $\mathcal{F} = \mathcal{F}_0$ for $|t| \leq \varepsilon$ is then given by submersions $\Phi^\alpha_t : U'_\alpha \to V$, related on $U'_{\alpha\beta} = U'_\alpha \cap U'_\beta$ by

(4.6) $$\Phi^\alpha_t(x) = \gamma^{\alpha\beta}_t(\Phi^\beta_t(x)) \circ \Phi^\beta_t(x).$$

Consider on U'_α the vector field ν^α along $f^\alpha = \Phi^\alpha_0$ given by

(4.7) $$\nu^\alpha(x) = \frac{d}{dt}\Big|_{t=0} \Phi^\alpha_t(x).$$

Differentiating (4.6) at $t = 0$ we obtain

(4.8) $$\nu^\alpha(x) = \dot{\gamma}^{\alpha\beta}_0(f^\beta(x)) + \gamma^{\alpha\beta}_*(f^\beta(x))\nu^\beta(x).$$

Here we have set $\gamma^{\alpha\beta}_* = (\gamma^{\alpha\beta}_0)_*$, where generally $g^{\alpha\beta}_t = (\gamma^{\alpha\beta}_t)_*$ denotes the derivative of $\gamma^{\alpha\beta}_t$. Note that $g^{\alpha\beta}_t$ is the transition function for the normal bundle Q_t of the foliation $\mathcal{F}_t$ on $U'_{\alpha\beta}$. Formula (4.8) expresses the fact that the $\nu^\alpha \in \Gamma(U'_\alpha, (f^\alpha)^*TV_\alpha)$ together define a section $\nu \in \Gamma Q$ only if $\dot{\gamma}^{\alpha\beta}_0 = 0$.

Given $\nu^\alpha \in \Gamma(U'_\alpha, (f^\alpha)^*TV_\alpha)$, one obtains locally a variation Φ^α_t of $f^\alpha = \Phi^\alpha_0$ by setting

(4.9) $$\Phi^\alpha_t(x) = \exp_{f^\alpha(x)}(t\nu^\alpha(x)) \quad \text{for} \quad x \in U_\alpha, \ |t| \leq \varepsilon$$

where $\varepsilon > 0$ is sufficiently small. The RHS is the endpoint of the geodesic segment in $V_\alpha \subset N$ starting at $f^\alpha(x)$ and determined by $t\nu^\alpha(x) \in T_{f^\alpha(x)}N$. This is the construction of Eells-Sampson. Clearly (4.7) holds for this variation, and by [ES] we have for the derivative $(\Phi^\alpha_t)_*$

(4.10) $$\frac{\nabla \partial}{\partial t}\Big|_{t=0} (\Phi^\alpha_t)_* = \nabla\nu^\alpha \in \Omega^1(U'_\alpha, Q).$$

Let now $\nu \in \Gamma Q$ and $\nu^\alpha = \nu|U'_\alpha$. Then we claim that the local variations Φ^α_t of f^α define a global variation of $\mathcal{F}$, i.e. the compatibility relation (4.6) holds for $x \in U'_{\alpha\beta}$.

First observe that with $\gamma_*^{\alpha\beta} = g^{\alpha\beta}$ we have

$$\nu^\alpha(x) = g^{\alpha\beta}(x)\nu^\beta(x).$$

Since $\gamma^{\alpha\beta}$ is a local isometry of N, it sends parametrized geodesic to parametrized geodesics, and we have

$$\exp_{\gamma^{\alpha\beta}f^\beta(x)}(t\gamma_*^{\alpha\beta}(f^\beta(x))\nu^\beta(x)) = \gamma^{\alpha\beta}\exp_{f^\beta(x)}(t\nu^\beta(x)).$$

It follows that

(4.11) $$\phi_t^\alpha(x) = \gamma^{\alpha\beta}\phi_t^\beta(x), \quad \text{as claimed.}$$

We observe that the variation F_t of F so defined for $\nu \in \Gamma Q$, $|t| \leq \varepsilon$ has for each t the same t-independent cocycle $\gamma^{\alpha\beta}$. We call these variations special variations associated to sections of Q. Since the cocycle $\gamma^{\alpha\beta}{}_* = g^{\alpha\beta}$ defines the normal bundle Q, and $\gamma^{\alpha\beta}$ defines the holonomy of the foliation, these data are unchanged under special variations.

Since g_Q is a holonomy invariant fiber metric on Q, and these data do not change, F_t is a Riemannian foliation for each t with respect to the same metric g_Q. But the metric g_M on M which is bundle-like for F_o has no reason to be bundle-like for F_t, $t \neq 0$. In particular the evaluation formula in Proposition 4.4 only holds for F_o. We formulate the main result of this section.

4.12 THEOREM. Let M be a compact oriented manifold, with a Riemannian foliation F and bundle-like metric g_M.

(i) Any $\nu \in \Gamma Q$ defines a special variation of F by the local definition (4.9). If $\pi_t : TM \to Q$ denotes the projection of F_t, then $\frac{\nabla\partial}{\partial t}\Big|_{t=0} \pi_t = \nabla\nu \in \Omega^1(M,Q)$.

(ii) F is harmonic if and only if it is an extremal of the energy functional for special variations of F.

Proof. (i) has been established before. The property $\dot{\pi}_o = \frac{\nabla\partial}{\partial t}\Big|_{t=0} \pi_t = \nabla\nu$ is (4.10). To establish (ii), we cna now calculate as in [ES]:

$$\frac{d}{dt}\Big|_{t=0} E(F_t) = \tfrac{1}{2}\frac{d}{dt}\Big|_{t=0} \|\pi_t\|^2 = \langle\dot{\pi}_o,\pi_o\rangle = \langle\nabla\nu,\pi\rangle = \langle d_\nabla\nu,\pi\rangle = \langle\nu,d_\nabla^*\pi\rangle.$$

Thus $\frac{d}{dt}\Big|_{t=0} E(F_t) = 0$ for all special variations ν of F is and only if the tension $\tau = d_\nabla^*\pi$ vanishes, i.e. if and only if F is harmonic.

5. COMMENTS

The theory of harmonic foliations here initiated is modelled on the Eells-Sampson theory of harmonic maps. Now a p-dimensional foliation on M can also be given by a section of the Grassmannian bundle of p-planes over M, its Gauss section. The harmonicity of the foliation can be expressed in terms of its Gauss section. The motivation for this is the result of Ruh-Vilms [RV] on the harmonicity of the Gauss map of a minimal submanifold $M \subset \mathbb{R}^n$.

The harmonic foliations enjoy certain regularity properties, as do the harmonic maps. Let M be a compact and smooth manifold, and a C^2-foliation on M, i.e. the local submersions are C^2-maps. Then the harmonicity of F with respect to a Riemannian metric g_M on M implies the smoothness of F. This can be sharpened in the same way as the corresponding statements for harmonic maps (see [EL, p. 10]).

An interesting question would be to determine all harmonic foliations of a given Riemannian manifold. For $\mathbb{R}^n$ there are certainly no harmonic foliations with compact leaves. What about harmonic foliations with complete leaves? Under which conditions is such a foliation equivalent to the coset foliation of $\mathbb{R}^n$ by a p-dimensional subspace? The next simplest case would be to consider harmonic foliations of the sphere S^n. The Hopf fibration $S^3 \to S^2$ is an example. Next one will have to turn to Riemannian homogeneous spaces. One expects many of the classical results on minimal submanifolds to reappear in a refined form.

The second variation formula for the energy of a harmonic foliation has interesting applications. Some of these are discussed in [KT7].

REFERENCES

[EL] J. Eells and L. Lemaire, A report on harmonic maps, Bull. Londo Math. Soc. 10(1978), 1-68.

[ES] J. Eells and J. H. Sampson, Harmonic mappings of Riemannian manifolds, Amer. J. Math. 86(1964), 109-160.

[EP1] D. B. A. Epstein, Foliations with all leaves compact, Ann. Inst Fourier 26(1976), 265-282.

[EP2] D. B. A. Epstein, Foliations with all leaves compact, Lecture Notes in Mathematics 468(1974), 1-2.

[ER] D. B. A. Epstein and H. Rosenberg, Stability of compact foliations, Lecture Notes in Mathematics 652(1978), 151-160.

[GV] C. Godbillon et J. Vey, Un invariant des feuilletages de codime sion un, C. R. Acad. Sc. Paris 273(1971), 92-95.

[HA] A. Haefliger, Some remarks on foliations with minimal leaves, to appear.

[HE] H. L. Heitsch, Independent variation of secondary classes, Annals of Math. 108(1978), 421-460.

[HM] R. S. Hamilton, Deformation theory of foliations, preprint Cornell University (1978).

[KN] S. Kobayashi and K. Nomizu, Foundations of differential geomet I, II (1963, 1969).

[KT1] F. W. Kamber and Ph. Tondeur, Invariant differential operators and the cohomology of Lie algebra sheaves, Memoirs Amer. Math. Soc. 113(1971), 1-125.

[KT2] F. W. Kamber and Ph. Tondeur, Characteristic invariants of fol ated bundles, Manuscripta Math. 11(1974), 51-89.

[KT3] F. W. Kamber and Ph. Tondeur, Foliated bundles and characteris classes, Lecture Notes in Mathematics 493 (1975).

[KT4] F. W. Kamber and Ph. Tondeur, Non-trivial characteristic invar iants of homogeneous foliated bundles, Ann. Scient. Ec. Norm. Sup. 8(1975), 433-486.

[KT5] F. W. Kamber and Ph. Tondeur, On the linear independence of certain cohomology classes of $B\Gamma$, Advances in Math. Suppl. Studies 5(1979), 213-263.

[KT6] F. W. Kamber and Ph. Tondeur, Feuilletages harmoniques, C. R. Acad. Sc. Paris 291(1980), 409-411.

[KT7] F. W. Kamber and Ph. Tondeur, Infinitesimal automorphisms and second variation of the energy for harmonic foliations, to app

[L] H. B. Lawson, Jr., Lectures on minimal submanifolds, Vol. I (1980), Publish or Perish, Inc.

[LP] C. Lazarov and J. Pasternak, Residues and characteristic class for Riemannian foliations, J. Diff. Geom. 11(1976), 599-612.

[ON] B. O'Neill, The fundamental equations of a submersion, Michigan Math. J. 13(1966), 459-469.

[P] J. Pasternack, Foliations and compact Lie group actions, Comment. Math. Helv. 46(1971), 467-477.

[RE1] B. L. Reinhart, Foliated manifolds with bundle-like metrics, Annals. of Math. 69(1959), 119-132.

[RE2] B. L. Reinhart, The second fundamental form of a plane field, J. Differential Geometry 12(1977), 619-627.

[RU1] H. Rummler, Quelques notions simples en géométrie riemannienne et leurs applications aux feuilletages compacts, Comment. Math. Helv. 54(1979), 224-239.

[RU2] H. Rummler, Kompakte Blätterungen durch Minimalflächen, Habilitationsschrift Universität Freiburg i. Ue. (1979).

[RV] E. Ruh and J. Vilms, The tension field of the Gauss map, Transactions Amer. Math. Soc. 149(1970), 569-573.

[S] D. Sullivan, A homological characterization of foliations consisting of minimal surfaces, Comment. Math. Helv. 54(1979), 218-223.

Department of Mathematics
University of Illinois at
Urbana-Champaign
Urbana, Illinois 61801

On the Stability of Harmonic Maps

Pui-Fai Leung

§ 0. Introduction

Throughout this paper we denote by N^m a m-dimensional compact Riemannian manifold without boundary. A smooth map $f: N^m \to M^n$ where M^n is an arbitrary n-dimensional Riemannian manifold is called harmonic if it is a critical point of the energy integral. A harmonic map is called stable if it has non-negative second variation. It is a result of R. T. Smith [3] that for $n \geq 3$, the index of the identity map on S^n is $n+1$ and so the identity map is unstable. Recently Y. L. Xin [4] proved that for $m \geq 3$, any stable harmonic map $f: S^m \to M^n$ must be constant where M^n can be any Riemannian manifold. In this paper we shall prove the following two theorems :-

Theorem A. Let M^n be a complete orientable hypersurface of $\mathbb{R}^{n+1}$ (= (n+1) - dimensional Euclidean space).
Denote the principal curvatures by $\lambda_1, \ldots, \lambda_n$ and let $\lambda^2 = \max\{\lambda_i^2\}_{i=1}^n$.
Let $\underline{K}$ be the function that assigns to each point in M the minimum of the sectional curvatures at that point.
Suppose that $\lambda^2 < (n-1)\underline{K}$.
Then any stable harmonic map $f: N^m \to M^n$ is a constant map.

Corollary 1 Any stable harmonic map from any N^m to S^n where $n \geq 3$ is a constant.

Proof: This follows from theorem A since for S^n in $\mathbb{R}^{n+1}$ we have $\lambda_1 = \ldots = \lambda_n = 1$ and $\underline{K} = 1$. //

Remark 1. If M^n is compact, then the condition $\lambda^2 < (n-1)\underline{K}$ implies that M^n is a homotopy n-sphere by a theorem of J. D. Moore [2].

Remark 2. During the Harmonic map conference at Tulane University we learned from Prof. J. Eells the following results :-

Theorem Let (M,g) be compact and (N,h) a n-sphere with $n \geq 3$ and any metric h. Then (a) any harmonic map $f: N \to M$ which is a minimum of energy in its homotopy class must be constant; (b) if $f: M \to N$ is a minimum of energy in its homotopy class and if that class contains a submersion, then f must be a constant.

We would like to thank Professor Eells for many discussions and encouragements during the conference.

Theorem B. Let M^n be a compact manifold without boundary isometrically immersed in some $\mathbb{R}^{n+p}$. Denote by ρ the scalar curvature of M, S the square length of the second fundamental form of the immersion. Suppose $S < \rho$, then the identity map on M^n is unstable.

Corollary 2 Let M^n be a compact manifold without boundary isometrically immersed in some S^{n+p}. If $S + n < \rho$, then the identity map on M^n is unstable.

Proof: This follows from theorem B since

$$S(M^n \text{ in } R^{n+p+1}) = S(M^n \text{ in } S^{n+p}) + n.$$

//

Corollary 3 Let M^n be a compact manifold without boundary isometrically and minimally immersed in some S^{n+p}.

If $\rho > \frac{1}{2} n^2$

then the identity map on M^n is unstable.

Proof: It follows from the equation of Gauss that $S = n(n-1) - \rho$ and hence result follows from corollary 2. //

Our proofs of the above theorems are based on a method first used by Lawson and Simons [1] and we would like to express our thanks to their paper.

§.1. Preliminaries

Consider a smooth map $f : N^m \to M^n$.

Denote by ∇, R, g(or < >) the Riemannian connection, the curvature and the metric tensor respectively on M^n.

The energy of f is defined by

$$E(f) = \frac{1}{2} \int_N \sum_{i=1}^{m} g(f_* e_i , f_* e_i) \, dv$$

where $\{e_1, \ldots, e_m\}$ is an orthonormal basis for TN and dV is the volume element on N.

For any vector field V on M, denote by ϕ_t the one-parameter group generated by V.

Consider the variation $f_t = \phi_t \circ f : N^m \to M^n$ of f. We write E(t) for $E(f_t)$, then we have

$$E^{(k)}(0) = \frac{1}{2}\int_N \sum_{i=1}^{m} \left((L_V^{(k)} g)(f_* e_i, f_* e_i)\right) dV$$

where L_V is the Lie derivative with respect to V. For any vector field V on M, we define two tensors α^V and $\nabla_{V,\cdot}V$ in Hom(TM,TM) corresponding to V by

$$\alpha^V(x) = \nabla_x V$$

$$\text{and} \quad \nabla_{V,x}V = \nabla_V \nabla_x V - \nabla_{\nabla_V x} V \quad .$$

We note that both extend naturally as derivations to the entire tensor algebra of M. We have the equation $\nabla_V - \alpha^V = L_V$ on the entire tensor algebra of M.

Using the above equation, the facts that $\nabla g = 0$ and $\alpha^V = 0$ on functions, we obtain after a direct computation that

$$E'(0) = \int_N \sum_i \langle \alpha^V(f_* e_i) , f_* e_i \rangle dV.$$

and $$E''(0) = \int_N \sum_i \{ \| \alpha^V(f_* e_i \|^2 + \langle (\alpha^V)^2 (f_* e_i), f \ e_i \rangle$$
$$+ \langle \nabla_{V, f_* e_i} V, f_* e_i \rangle \} dV$$

We note that

$$\nabla_{V,x} V = \alpha^{\alpha^V(V)}(x) - (\alpha^V)^2(x) + R(V,x)V.$$

and so if f is harmonic, we have

$$E''(0) = \int_N \sum_i \{ \| \mathfrak{a}^V(f_* e_i) \|^2 + \langle R(V, f_* e_i)V, f_* e_i \rangle \} \, dV$$

§ 2. Proof of theorems

We now consider the situation as in theorem A.

Let V be a fixed vector field in $\mathbb{R}^{n+1}$.

We denote by T,N the tangential and normal projection of vectors in $\mathbb{R}^{n+1}$ to M^n. We consider the second variation of a harmonic map f: $N^m \to M^n$ corresponding to V^T. We have

$$E''(0) = \int_N \sum_i \{ \| \mathfrak{a}^{V^T}(f_* e_i) \|^2 + \langle R(V^T, f_* e_i)V^T, f_* e_i \rangle \} \, dV$$

now

$$\begin{aligned} \mathfrak{a}^{V^T}(f_* e_i) &= \nabla_{f_* e_i}(V^T) \\ &= (\tilde{\nabla}_{f_* e_i}(\overset{0}{\not V} - V^N)^T \\ &= (-\tilde{\nabla}_{f_* e_i} V^N)^T \\ &= A^{V^N}(f_* e_i) \end{aligned}$$

where $\tilde{\nabla}$ is the connection on $\mathbb{R}^{n+1}$ and A^{V^N} is the second fundamental form corresponding to V^N:

Next, we have

$$\begin{aligned} &- \langle R(V^T, f_* e_i)V^T, f_* e_i \rangle \\ &= K(V^T, f_* e_i) \, \| V^T \wedge f_* e_i \|^2 \\ &= K(V^T, f_* e_i) \{ \| V^T \|^2 \| f_* e_i \|^2 - \langle V^T, f_* e_i \rangle^2 \} \end{aligned}$$

where $K(V^T, f_* e_i)$ is the sectional curvature on M corresponding to the plane spanned by V^T and $f_* e_i$.

Now we define a quadratic form Q on $\mathbb{R}^{n+1}$ by $Q(V) = E''(0)$ where $E''(0)$ is the second variation of f corresponding to V^T. We write $Q(V) = \int_N q(v)\,dv$ and we wish to calculate Trace (Q). We first note that at a point $y = f(x) \in M^n$, the value of Trace (q) at y is independent of the choice of orthonormal basis of $T\mathbb{R}_y^{n+1} \cong \mathbb{R}^{n+1}$ because q is a quadratic form on $T\mathbb{R}_y^{n+1}$. Now at the point y we choose an orthonormal basis $\{v_1, \ldots, v_{n+1}\}$ of $\mathbb{R}^{n+1}$ such that $v_1, \ldots, v_n$ are tangent to M^n at y and v_{n+1} is normal to M^n at y.

Then we have

$$\mathrm{Trace}(Q) = \int_N \{\sum_i \|A(f_* e_i)\|^2 - \sum_{j=1}^{n} \sum_{i=1}^{m} K(v_j, f_* e_i)\|v_j \wedge f_* e_i\|^2\}\,dV$$

where A is the second fundamental form of the hypersurface M^n in R^{n+1}.

We note that

$$\|A(f_* e_i)\|^2 \le \lambda^2 \|f_* e_i\|^2$$

and

$$K(v_j, f_* e_i)\|v_j \wedge f_* e_i\|^2 \ge \underline{K}\,\|v_j \wedge f_* e_i\|^2$$

hence

$$\mathrm{Trace}(Q) \le \int_N \{\sum_i \lambda^2 \|f_* e_i\|^2 - \underline{K} \sum_i \left(n\|f_* e_i\|^2 - \|f_* e_i\|^2\right)\}\,dV$$

$$= \int_N \{(\lambda^2 - \underline{K}(n-1)) \sum_i \|f_* e_i\|^2\}\,dV$$

Since f is stable we must have Trace $(Q) \geq 0$ but by assumption we have $\lambda^2 < (n-1)\underline{K}$, hence we must have $f_* \equiv 0$ and so f is a constant.

This completes the proof of theorem A. Theorem B is proved similarly.

Acknowledgement. The author would like to thank his supervisor, Professor T. Nagano, for much advice and his friend, J. H. Cheng, for many discussions. He would also like to thank Professors J. Eells, L. Lemaire and the referee for pointing out errors in the first version.

Added Remark. In theorem A, we assume M has positive sectional curvature and order the principal curvatures (with respect to a suitably chosen normal direction) so that $0 < \lambda_1 \leq \lambda_2 \leq \cdot \cdots \leq \lambda_n$, then $\underline{K} = \lambda_1 \lambda_2$ and the condition in theorem A can be written as $\lambda_n^2 < (n-1)\lambda_1 \lambda_2$. Using this condition and a direct computation of principal curvatures, we obtain the following class of examples:

Example. Let M^n be the ellipsoid in $\mathbb{R}^{n+1}$ given by the equation $Gx_1^2 + x_2^2 + \ldots + x_{n+1}^2 = 1$ and suppose $\frac{1}{n-1} < G < \sqrt{n-1}$, then any stable harmonic map $f : N^m \to M^n$ must be constant.

References

[1] Lawson & Simons: On stable currents and their application to global problems in real and complex geometry. Ann. of Math. 98 (1973) pp. 427-450.

[2] J. D. Moore: Codimension two submanifolds of positive curvature. Proc. of AMS 70 (1978) pp. 72-74.

[3] R. T. Smith: The second variation formula for harmonic mappings. Proc. of AMS 47 (1975) pp. 229-236.

[4] Y. L. Xin: Some results on stable harmonic maps. Duke Math. J. 47 (1980) pp. 609-613.

Department of Mathematics
University of Notre Dame
Notre Dame, Indiana 46556

Stability of Harmonic Maps Between Symmetric Spaces

T. Nagano*

Introduction. The question of stability of a harmonic map amounts to finding the first several eigenvalues of a certain symmetric elliptic linear differential operator, L , which appears in the second variation formula for the map. In general this is a difficult task because of the lack of appropriate estimates of the eigenvalues (despite of many efforts). In the presence of an abundant group action however, the problem can be easier to solve. And this note is to illustrate it with totally geodesic submersions of symmetric spaces onto others. The results were essentially obtained by R. T. Smith [10] by using [6]; it is a part of the purpose to make them more complete and add some geometric flavor (4.1). The same idea in Section 1 will be applied to another problem elsewhere.

Table of Contents.

1. The operator L or the problem setting. Let f be a harmonic map [3] of a compact Riemann manifold M into another. We say f is stable if its energy integral, $E(f) = (1/2)\int_M \|df\|^2$, is a local minimum of $E(f_t)$ for any variation or smooth homotopy (f_t) of $f = f_0$. When the identity map of M is stable, we say that M is stable. This is the case if and only if the second variation

(1.1) $$(d^2/dt^2)_{t=0}E(f_t) = \int_M \|\nabla v\|^2 - R(v,v)$$

---

* Partially supported by NSF Grant MCS 80 03573.

is non-negative for every vector field $v = (\partial/\partial t)_{t=0} f_t$ where R is the Ricci form (cf. [10]). In fact the integral (1.1) equals $\int_M \langle Lv, v\rangle$ for a symmetric elliptic linear differential operator L since

$$\int_M \|\nabla v\|^2 = \int_M (-g_{ab}) g^{jk} v^a \nabla_i \nabla_k v^b = \int_M \langle \Delta v, v\rangle + R(v,v)$$

by Stokes' theorem, where Δv is the Laplacian of the dual one-form to v, and consequently $\mathcal{T}M$, the Hilbert space of the vector fields with the norm by $\int_M \|v\|^2$, is the Hilbert sum of the finite dimensional eigenspaces of L.

There is a particularly fortunate case in which we have a definite answer.

(1.2) M <u>is stable if</u> M <u>is Kaehlerian</u>.

<u>Proof</u>. Express the maps with their graphs. The graph of the identity map is the diagonal of $M \times M$, which is a totally geodesic and hence minimal submanifold. Since the second fundamental form of the diagonal is zero, its second variation for the energy integral up to a constant multiple (see the remark (1.4) below.) Since every compact complex submanifold of a Kaehlerian manifold is stable as a minimal submanifold by Federer's theorem [4], it follows immediately that M is stable.

Q.E.D.

<u>Remark</u>. Actually Smith [10] proved that every holomorphic map is stable. We obtained the above proof during conversation with A. Howard. From the proof, we obtain the following.

(1.22) <u>A compact Riemann manifold</u> M <u>is stable if and only if the diagonal in</u> $M \times M$ <u>is stable as a minimal submanifold</u>.

Now let us assume that the connected isometry group G of M is transitive, so $M = G/K$. G acts on $\mathcal{T}M$ and its elements commute with L, since L is constructed by using the metric only. Thus each eigenspace of L is left invariant by G and is the direct sum of G-invariant irreducible subspaces. Let V be one. Then we have the <u>Casimir operator</u> $C|V$ which is defined as follows. We fix an orthonormal basis (e_λ) for the Lie algebra $\mathfrak{L}(G)$ of the compact group G with respect to a

G-invariant metric, which is compatible with the metric of M. Let $\rho(e_\lambda)$ denote the representation or the action of e_λ on V. In the present case, $\rho(e_\lambda) = [e_\lambda, v$ if e_λ is understood as a vector field on M. $C|V$ is by definition the operator $(-1)\Sigma_\lambda \rho(e_\lambda)^2$. $C|V$ is a scalar multiple of the identity ([1], Chap. 8), and this scalar can be calculated explicitly if one knows the action of G on V and uses Freudenthal's formula ([1], Chap. 8, p. 120). Every such V is known with the following rules. A G-module V appears in $\mathcal{T}M$ if and only if the K-module TM_x is isomorphic with a submodule the K-module V , $k(x) = x$. Besides the G-module V appears in $\mathcal{T}M$ as many times as TM_x does in the k-module V. (These rules constitute the Frobenius theorem as formulated by R. Bott.) So we know the eigenspaces together with the eigenvalues of the symmetric elliptic linear differential operator C , as principle. $\mathcal{T}M$ is then the Hilbert sum of the common eigenspaces of C and L. The Lie algebra $\mathcal{L}(G)$ is one of them, on which L is zero and C may be normalized to be the identity, provided G is semisimple. Thus one can determine whether or not M is stable by routine computations of the possible eigenvalues < 1 of C which can appear in $\mathcal{T}M$, if the order (<) of the eigenvalues of C matches the order of those of L on the common eigenspaces. I don't know if this last condition is met in general and can be proven by considering the scalar multiplication of functions by V. But this is the case for the stability of the symmetric spaces.

Our problem of stability has a nice feature: we have only to consider the gradients of functions on M by the following formula ((1.11) in [11]).

$$\text{(1.3)} \qquad \int_M ||\nabla v||^2 - R(v,v) = \int_M (1/2)\, ||\mathcal{L}_v g||^2 - (\mathrm{div}(v))^2 .$$

The Casimir operator on the space of the functions is the Laplacian.

(1.4) Remark. To conclude this section, we like to explain the second variation formula for the volume integral. Let f be a map of a compact Riemann manifold M into another. Then $F := (df)^* \circ df$, the composite with the dual, is an endomorphism of the tangent bundle TM. We have the function $p_k = p_k(f)$ which assigns the k-th elementary symmetric polynomial of the eigenvalues of F to each point of M. We may consider the integral $\int_M p_k$ or more generally the integral of any reasonable function, Φ, of $p_1, p_2, \ldots, p_n$, $n = \dim M$. For instance, $p_1/2$ and $(p_n)^{1/2}$ give the energy integral and the volume integral respectively. When f is an isometric immersion, the condition of f being critical is independent of Φ. But the second variation formula varies with Φ ; for instance,

$$(1.5) \quad (d^2/dt^2)_{t=0} \int_M p_k(f_t) = 2 \int_M \binom{n-1}{k-1} \{\|dv\|^2 - R'(v,v]\} - 2\binom{n-2}{k-2} \|Bv\|^2$$

if v is normal to f(M) , where B is the second fundamental form of f(M). In case $B = 0$ as in (1.2) or (1.2a), the second variation formulas coincide again (up to a constant multiple). I have added this remark with the hope that someone might define an integral which fits the (Palais-Smale) Morse theory better for studying the space of mappings.

2. Symmetric spaces. A Riemann manifold M is called symmetric (see [5]) if there is a symmetry s_x of M for each point x which is locally described as $s_x(y) = -y$ with respect to the normal coordinates centered at x. The morphisms in the category of the symmetric spaces are those maps f which commute with the symmetries; thus $s_{f(x)} \circ f = f \circ s_x$ for every x in M. A map f is a morphism if and only if f is totally geodesic; i.e. f carries the geodesics into geodesics with the parameters counted. From now on we assume that the morphism f is a submersion of a compact symmetric space M onto another. Then f is a fibration and M is a local Riemann product by Proposition 6.7 on 6.7 in [7] since the metric of M is "bundle-like" or so it can be renormalized. We may assume without losing interest that M is irreducible and hence f is a covering map; i.e. M is not a local Riemann product. From the definition it follows easily that the symmetries generate a transitive isometry group of M , if M is connected. Let G denote the identity component of the isometry group of M ; thus $M = G/K$. The action of K on the tangent space TM_x , $K(x) = x$, is irreducible if and only if M is irreducible (modul minor details). It is also a direct consequence of the definition that for each point $\propto$ there is an orthonormal basis (e_λ) for the Lie algebra

$L(A)$ such that $e_1(x), \ldots, e_n(x)$ span TM_x and $e_i = 0$ at x, $1 < i < n$. This implies that $C - L$ is a differential operator of order zero; i.e. the difference of the Casimir operator C on TM and the operator L defined in Section 1 is a symmetric linear endomorphism, P, of TM. Since G commutes with P obviously, P is a constant multiple of the identity by Schur's lemma. Therefore the idea explained in Section 1 works for the study of stability of compact symmetric spaces; in particular, (2.1) M is stable if and only if the Casimir operator on TM or on the functions has no eigenvalue < 1.

As to the representation of G in the space of functions, FM, we have the following special case of the Bott-Frobenius theorem.

(2.2) An irreducible representation on space V of G appears in FM, the space of functions, if and only if K fixes a nonzero vector of V.

It is not trivial but true that, when V appears in FM, V does just once.

3. Group manifolds. Every compact Lie group M is a symmetric space with respect to a bi-invariant Riemannian metric (by the symmetry $s_1(x) = x^{-1}$ at the unit 1). Let M be a compact connected (semisimple) Lie group. Then $G = M \times M$ is the right and the left translation group. K is the diagonal. An irreducible representation $\rho \otimes \sigma$ of M appears in FM if and only if ρ is dual to σ, as one sees from (2.2). One may exploit the aforementioned formula of Freudenthal to conclude the following.

(3.1) A compact, simple, 1-connected Lie group M is stable if and only if M is the exceptional group E_8.

(3.2) A compact simple connected Lie group M with trivial center is stable if and only if M is the adjoint group of $SU(n)$, $SO(2n)$, E_6, E_7 or E_8.

We will record more precise information, which enables one, say, to calculate the index of L, or the total multiplicity of the negative eigenvalues of L. We follow Bourbaki's notation [1] to denote the fundamental representations. The irreducible representations, ρ above, of the 1-connected M whose Casimir operators have eigenvalues < 1 (corresponding to the adjoint representation) are $\tilde{\omega}_1$, $\tilde{\omega}_2$, $\tilde{\omega}_{n-2}$, and $\tilde{\omega}_{n-1}$ for $SU(n)$; $\tilde{\omega}_1$ for $Spin(n)$, $\tilde{\omega}_n$ for $Spin(2n+1)$ with $n = 3$ or 4; $\tilde{\omega}_1$ and $\tilde{\omega}_2$ for $Sp(n)$ and $\tilde{\omega}_3$ for $Sp(3)$; $\tilde{\omega}_n$ and $\tilde{\omega}_{n-1}$ for $Spin(2n)$ with $3 \le n \le 7$; $\tilde{\omega}_1$ and $\tilde{\omega}_6$ for E_6; $\tilde{\omega}_7$ for E_7; $\tilde{\omega}_4$ for F_4; and $\tilde{\omega}_1$ for G_2; where $Spin(n)$ is the universal covering group of $SO(n)$.

4. Geometry of unstable manifolds. We consider the compact 1-connected symmetric spaces $M = G/K$ with simple G. M is then irreducible. R. T. Smith [10] found for the classical spaces that M is unstable if and only if M is the sphere S^n, $n \ge 3$, or $SU(2n)/Sp(n)$. From the exceptional spaces we should add the Cayley projective plane $F_4/Spin(9)$. Instead of using [6], one could find these in a more geometric way to see the following theorem.

(4.1) Let $M = G/K$ be an unstable 1-connected symmetric space with compact and simple G. Then M has the following properties:

(a) M admits a totally geodesic immersion i into another symmetric space $M' = G'/K'$ such that (i) $K' = G$ or $SO(2) \times G$ and (ii) $i(M)$ is a connected component of the fixed point set of M'_+ of the symmetry s_x at some point x of M' or this is true if one projects $i(M)$ into a symmetric space of which M' is a covering space;

(b) The non-compact form G'' of G' with respect to G'/M' can act on M and as such contains G properly;

(c) The submanifold $i(M)$ is unstable as a minimal submanifold of M'.

Comments on (4.1). First we explain (4.1) in the case of the sphere S^n. (a) says in this case that S^n is embedded in S^{n+1} as the equator. If we project S^{n+1} onto the real projective space, then the image of the equator $i(M)$ is a connected component of the fixed point set of the symmetry at the dual point to the image hyperplane. (b) says the connected isometry group $G = SO(n+1)$ is contained in the conformal transformation group of S^n. Finally (c) says that the equator $i(S^n)$ of S^{n+1} is unstable as a minimal submanifold. (Note however that the projection of $i(S^n)$ in the projective space is stable.) Turning to the general case, one is lead to the immersion by (2.2), which is equivalent to say that one has a G-equivariant map, i , of M into V. Since M is irreducible, i is an immersion. Since the immersion is G-equivariant, it is totally geodesic. The property (ii) is read off immediately from a table in [2] which lists M_+. The case $K' = SO(2) \times G$ occurs when $M = SU(2n)/Sp(n)$. In this case $U(2n)/Sp(n)$ is a connected component of M'_+ for $M' = SO(4n)/U(n)$. Perhaps one can prove (b) *a priori*, but a look at Section 5 of [8] makes (b) evident. To prove (c), one may proceed along the line of Section 1 with TM replaced by the normal bundle. A crucial point is the existence of a G-invariant unit normal vector field (for which the Casimir operator is certainly zero). For $M = Sp(p+q)/(Sp(p) \times Sp(q))$ and $F_4/Spin(9)$, M' are $SU(2p + 2q)/Sp(p + q)$ and E_6/F_4 respectively.

(4.2) *The map* i *in* (4.1) *is homotopic to a constant map; in particular the image* $i(M)$ *is homologous to zero.*

Proof. This comes from the fact [2] that the points in M'_+ are characterized as the antipodal points of x. (See (a) in (4.1)) on closed circles of the circle for each point of $i(M)$. Then one can deform $i(M)$ along these geodesics to x.

Q.E.D.

Finally a small remark on the nullity, $\nu(M)$, which is $\dim Ker(L)$. From [9] one obtains

(4.3) *The nullity* $\nu(M) = \dim G$ *except that* $\nu(M) = 2 \dim G$ *for Kaehlerian* M *and* $\nu(G_2) > \dim G$, *if* M *is a compact connected irreducible symmetric space.*

BIBLIOGRAPHY

[1] N. Bourbaki, Groupes et algebres de Lie, Hermann, 1975.

[2] B. Y. Chen - T. Nagano, Totally geodesic submanifolds of symmetric spaces II. Duke Math. J. 45 (1978), 405-425.

[3] J. Eells - L. Lemaire, A report on harmonic maps. Bull. London Math. Soc. 10 (1978), 1-68.

[4] H. Federer, Some theorems on integral currents. Trans. Am. Math. Soc. 117 (1965), 43-67.

[5] S. Helgason, Differential geometry, Lie groups and symmetric spaces. Acad. Press, 1978.

[6] T. Nagano, On the minimal eigenvalues of the Laplacians in Riemannian manifolds. Sci. Pop. Coll. Gen. Ed., Univ. of Tokyo, 11 (1961), 177-182.

[7] T. Nagano, On fibred Riemannian manifolds. Sci. Pop. Coll. Gen. Ed., Univ. of Tokoyo 10 (1960), 17-27.

[8] T. Nagano, Transformation groups on compact symmetric spaces. Trans. Am. Math. Soc. 118 (1965), 428-453.

[9] T. Nagano - K. Yano, Les champs des vecteurs geodesiques sur les espaces symetriques. C. R. Paris 252 (1961), 504-505.

[10] R. T. Smith, The second variation formula for harmonic maps. Proc. Am. Math. Soc. 47 (1975), 229-236.

[11] K. Yano, Integral formulas in Riemannian geometry. Dekker, 1970.

University of Notre Dame
Notre Dame, Indiana 46556

ON A CLASS OF HARMONIC MAPS

by

J. H. Sampson*

1. Here we shall describe an extensive class of harmonic maps. They are in fact <u>totally geodesic</u> maps, meaning that the associated second fundamental form vanishes. For a mapping $f : M \to Y$ of Riemannian manifolds the second fundamental form is the field given in local coordinates ($ds^2 = g_{ij}\,dx^i dx^j$ resp. $ds'^2 = g'_{\alpha\beta}dy^\alpha dy^\beta$) by the components

$$y^\alpha_{i|j} = y^\alpha_{ij} - \Gamma^k_{ij}\, y^\alpha_k + \Gamma'^\alpha_{\beta\gamma}\, y^\beta_i y^\gamma_j ,$$

where $y^\alpha_i = \partial y^\alpha/\partial x^i$ and $y_{ij} = \partial^2 y^\alpha/\partial x^i \partial x^j$. The trace, $g^{ij}\, y^\alpha_{i|j}$ gives the tension field τf.

2. Let $M = \Gamma\backslash G/K$ be a locally symmetric Riemannian manifold, where K is compact and Γ is discrete and operates without fixed points on G/K. Let $\rho : G \to G' = GL_m R$ (resp. $GL_m C$) be a linear representation of the Lie group G. If K' is a compact subgroup of G' which contains $\rho(K)$, then ρ induces a mapping

$$f : G/K \longrightarrow G'/K' .$$

If moreover $\rho(\Gamma) = \Gamma'$ is a discrete subgroup of G', then f induces

$$\bar{f} : \Gamma\backslash G/K \longrightarrow \Gamma'\backslash G'/K' .$$

It is easily seen that f is totally geodesic. For if x and y are canonical coordinates at the identity elements of G, G', then via ρ the y must be linear functions of the x^i (recall that for Lie groups one has $\exp d\rho(X) = \rho(\exp X)$). The Christoffel symbols vanish at the identity elements, and it is then clear that $y^\alpha_{i|j} = 0$ there, hence everywhere, by translation. Thus ρ is totally geodesic (cf. [1], p. 118). Now it is known [2, Chapter 4] that the x-system can be chosen so that a suitable part of it gives a normal coordinate system at the origin of G/K, and similarly for the y-system. It follows readily that f is totally geodesic.

3. The conditions given above are often realized in the case of Grassmann varieties, as we now indicate. For the n-sphere SO_{n+1}/SO_n and a representation $\rho : SO_{n+1} \to SO_{m+1}$, the subgroup $\rho(SO_n)$ will certainly be contained in a suitable choice of $SO_m \subset SO_{m+1}$ since SO_n is simple except for $n = 2, 4$ -- cases easily handled.

Thus we obtain totally geodesic maps $S^n \to S^m$. For the complex Grassmannian $G(p,p+q) = U_{p+q}/U_p \times U_q$, let $\rho : U_{p+q} \to U_N$ be a representation. Then the subgroup $\rho(U_p \times U_q)$ of U_N can in various ways be embedded in a subgroup of the form $U_P \times U_Q$ $(P + Q = N)$, giving us harmonic maps $G(p,p+q) \to G(P,P+Q)$. Many similar examples can be given.

---

*Supported in part by a grant from the National Science Foundation.

References:

1. James Eells and J.H. Sampson, Harmonic Mappings of Riemannian manifolds, American Jour. of Mathematics 86 (1964), 109-160.

2. S. Helgason, Differential Geometry and Symmetric Spaces, Academic Press, New York, 1962.

HARMONIC DIFFEOMORPHISMS OF SURFACES

H. C. J. Sealey

1. In [10] Shibata addressed himself to the following problem: Let M and N be two homeomorphic closed Riemann surfaces and h a conformal metric on N. A homotopy class of homeomorphisms $\phi : M \to N$ is given. Find a homeomorphism in that class which extremises the energy

$$E(\phi) = \int_M (|w_z|^2 + |w_{\bar{z}}|^2)\sigma^2(w)\,dxdy.$$

Here $z = x + iy$ is a complex coordinate on M, w is a complex coordinate on H, $h = \sigma^2(w)dwd\bar{w}$ and ϕ is represented by $z \mapsto w(z)$.

Shibata was interested in the case when the metric h is continuous and for that case a solution to the problem must satisfy a certain non-linear equation which reads:

$$\sigma^2(w)w_z\bar{w}_z \quad \text{is holomorphic} \qquad (*)$$

Let's say a map satisfying (*) is S-harmonic.

Now suppose that h is of class C^∞, and hence so is $\sigma(w)$, w is continuous and has locally square integrable distributional derivatives which are those appearing in (*). A natural question to ask, then, is:

Does the map ϕ have to be of class C^∞ ?

A positive answer to this question together with an existence theorem would then establish the existence of a harmonic diffeomorphism in the prescribed homotopy class of maps. At present very little is known about this problem. However, there are some partial results which I propose to bring together in this article.

2. The following positive result may be found in Sampson [6] and Schoen-Yau [7].

Theorem F. If the metric h has non-positive Gaussian curvature, then any harmonic map $\phi : M \to (N,h)$ of degree 1 is a diffeomorphism.

This fact, combined with the general existence theorem of Eells-Sampson [3] provides an existence theorem for harmonic diffeomorphisms in this special case.

3. As explained above, the question of existence of a harmonic diffeomorphism falls into two parts. (a) does a candidate exist? and (b) is this candidate a smooth map?

In his paper [1], packed full of ideas, Shibata claimed to have answered (a) affirmatively. However, it came to light that there were several gaps in his proof some of which have been patched but others still remain. For example, the following fact, essential to Shibata's argument, was established by Lelong-Ferrard [4] some few years later.

Theorem. Let M, N be closed Riemann surfaces and suppose N is equipped with a conformal metric. The set of all homeomorphisms $\phi : M \to N$ with $E(\phi)$ uniformly bounded is uniformly equicontinuous.

This fact has allowed the author [8] to establish the existence of a sequence of homeomorphisms converging uniformly to a map which is S-harmonic. The idea is to study a perturbed energy integral

$$E_\varepsilon(\phi) = \int_M (|w_z|^2 + |w_z|^2)\sigma^2(w)\,dxdy + \varepsilon\int_M \frac{|w_z|^2 + |w_z|^2}{|w_z|^2 - |w_z|^2}\rho^2(z)\,dxdy$$

where $\rho^2(z)dzdz$ is some arbitrary but fixed metric on M. For homeomorphisms whose inverses have square integrable derivatives, the second term is the energy of the inverse map. For $\varepsilon > 0$, the functional is easily minimized among all such maps. For any suitable map define

$$a(z) = \sigma^2(w)w_z\bar{w}_z \ , \quad k(z) = \frac{|w_z|^2 + |w_{\bar{z}}|^2}{|w_z|^2 - |w_{\bar{z}}|^2}$$

and

$$\rho(z) = \frac{\rho^2(z)w_z w_z}{|w_z|^2 - |w_z|^2}$$

so that k is a globally defined function on M while a and describe quadratic differentials on M.

Proposition. [8]

The minimizing map ϕ_ε satisfies

$$(a+\varepsilon)\lambda_z \, dxdy = \frac{\varepsilon}{2}\int k(\lambda^2)_z \, dxdy \qquad (**)$$

for all smooth functions λ with support in a coordinate chart.

The argument proceeds by studying (**) as $\varepsilon \to 0$ and showing that the limiting equation is (*).

4. A solution to (a) has, in fact, been given. Recall [1] that a K-quasiconformal map is a solution to Beltrani's equation

$$w_{\bar{z}} = \mu w_z \quad \text{where} \quad \text{ess sup}|\mu| = k < 1, \; K = \frac{1+k}{1-k}.$$

The following is due to Seratov [9].

Theorem 3.

For each sufficiently large K any homotopy class of K-quasiconformal maps from M to N contains a map satisfying (*).

Sketch of Proof.

The class of maps described is easily shown to contain a map minimizing E. Assume for a contradiction that (*) is violated so that $\frac{\partial a}{\partial \bar{z}} \not\equiv 0$ in the distributional sense.

Using the Hahn-Banach extension theorem it is possible to construct a Beltrami coefficient ν so that

$$\int_M a\nu dxdy = \int a\mu dxdy \quad \text{where} \quad \mu = \frac{w_{\bar{z}}}{w_z} \quad \text{a.e.}$$

while $\int_M \psi\nu \, dxdy = 0$ whenever ψ is holomorphic. This last fact means there is a variation of M given by

$$z' = H(z,\varepsilon) = z + \varepsilon h(z)$$

with $h_{\bar{z}} = \nu$. Defining $\phi_\varepsilon(z) = \phi(H(z,\varepsilon))$ the well-known formula for the first variation of E, see [1], [2] or [8],

$$\left.\frac{dE(\phi_\varepsilon)}{d\varepsilon}\right|_{\varepsilon=0} = \int_M a h_{\bar{z}} \, dxdy = \int_M a\nu dxdy = \int_M a\mu dxdy$$

$$= \int_M \sigma^2(w) w_z w_z \cdot \frac{w_{\bar z}}{w_z}\, dxdy$$

$$= \int_M \sigma^2(w) |w_{\bar z}|^2\, dxdy > 0$$

since the map ϕ is not conformal. Of course the hard work of the proof is showing that ϕ_ε is K-quasiconformal for small ε. A reference to this calculation may be found in [9].

5. The question now remaining to be resolved is whether a quasi-conformal map as described in Section 4 is smooth when the metric $h = \sigma^2(u)dwd\bar{w}$ is smooth and positive.

Example [1]

A Teichmüller map $T : M \to N$ is S-harmonic for a metric $h = |\psi|$ where ψ is a certain holomorphic quadratic differential on N. Away from the zeros of ψ, T is very well-behaved; in fact it is in suitable complex coordinates. At the zeros of ψ it is not even differentiable.

The best result to date is the following [8].

Proposition.

Suppose $\phi : M \to (N,h)$ is an S-harmonic C' diffeomorphism. Then ϕ is a harmonic diffeomorphism.

Proof. The content of the statement is that ϕ is C^∞.

For a smooth map $\psi : M \to N$ and a choice of smooth metric $g = \rho^2(z)dzd\bar{z}$ on M

$$\tau(\psi)^{1,0} = \frac{-4}{\sigma^2(w)\rho^2(z)} \cdot \frac{w_{\bar z}a_{\bar z} - w_z\bar{a}_z}{|w_z|^2 - |w_{\bar z}|^2}$$

as can be seen from a direct computation. Consequently, if ψ_ε is a variation of ψ with $\left.\frac{d\psi_\varepsilon}{d\varepsilon}\right|_{\varepsilon=0} = v$, then

$$\left.\frac{d}{d\varepsilon} E(\psi_\varepsilon)\right|_{\varepsilon=0} = \int_M \langle d\psi, \nabla v\rangle * 1 = -\int_M \langle \tau(\psi), v\rangle * 1$$

so that $\int_M \langle d\psi, \nabla v\rangle * 1 = 4\mathrm{Re} \int_M \left(\frac{w_{\bar z}a_{\bar z} - w_z\bar{a}_z}{|w_z|^2 - |w_{\bar z}|^2}\right) \bar{v}dxdy$. If now ψ_n is a sequence of smooth diffeomorphisms approximating ϕ in the C^1 topology, then, with the obvious notations,

$$\int_M \langle d\phi, \nabla v\rangle * 1 = \lim_{n\to\infty} \int_M \langle d\psi_n, \nabla v\rangle * 1$$

$$= \lim_{n\to\infty} 4 \operatorname{Re} \int_M \left(\frac{\partial w_n}{\partial \bar z} \cdot \frac{\partial a_n}{\partial \bar z} - \frac{\partial w_n}{\partial z} \frac{\partial \bar a_n}{\partial z} \right) \frac{\bar v}{J_n} \, dxdy$$

where $J_n = \left|\frac{\partial w_n}{\partial z}\right|^2 - \left|\frac{\partial w_n}{\partial z}\right|^2$. Thus

$$\int_M \langle d\phi, \nabla v\rangle * 1 = 4 \operatorname{Re} \int_M \frac{w_{\bar z} a_z - w_z \bar a_z}{|w_z|^2 - |w_{\bar z}|^2} \, dxdy$$

by the C^1 convergence $\psi_n \to \phi$. Since ϕ is S-harmonic $a_{\bar z} = 0$ so

$$\int_M \langle d\phi, \nabla v\rangle * 1 = 0.$$

Since v can be an arbitrary infinitesimal variation of ϕ, ϕ is a C^1 critical part of E. In particular ϕ is smooth as follows from [5] theorem 1.10.6.

To conclude, the problem which must be answered is the following. Suppose $\phi : M \to (N,h)$ is an S-harmonic quasi-conformal map. If h is smooth strictly positive metric on N, does it follow that ϕ is smooth?

A positive answer to this question would prove the existence of a harmonic diffeomorphism in the prescribed homotopy class.

REFERENCES

1. Ahlfors, L. V., On quasiconformal mappings, J. d'Analyze Math. 4 (1954), 1-58.
2. Courant, R., Dirichlet's Principle, conformal mappings and minimal surfaces, Interscience 1950.
3. Eells, J. and Sampson, J. H., Harmonic mappings of Riemannian manifolds, Amer. J. Math., 86 (1964), 109-160.
4. Lelong-Ferrard, J., Construction de modules de continuité dans le cas limite de Soboleft et applications à la geometrie différentielle, Arch. Rat. Mech. Anal., 52 (1973), 297-311.

5. Morrey, C. B., Multiple Integrals in the calculus of variations (Springer, 1966).

6. Sampson, J. H., Some properties and applications of harmonic mappings, Ann. Sci. Ec. Nom Sup. XI (1978), 211-228.

7. Schoen, R. and Yau, S. T., On univalent harmonic maps between surfaces, Invent. Math., 44 (1978), 265-278.

8. Sealey, H. C. J., Some properties of harmonic mappings, Thesis, University of Warwick, 1980.

9. Seretov , V. G., Functionals of Dirichlet type and harmonic quasi-conformal mappings, Soviet Math. Dokl., 14 (1973), 551-554.

0. Shibata, K., On the existence of a harmonic mapping, Osaka Math. J., 15 (1963), 173-211.

Equivariant Harmonic Maps into Spheres

Karen K. Uhlenbeck

Introduction:

Often interesting examples of solutions to non-linear problems are found by examining an equivariant case. In this article we examine the equations for equivariant harmonic maps $s: M \to S^k \subset R^{k+1}$, where $M = N \times R$ and $N = G/G_0$ is a compact symmetric space. We assume we have a representation ρ of the Lie group G in $SO(k+1)$ and $\rho(g)s(p) = s(gp)$ for all $g \in G$ and $p \in N \times R$.

The equations are similiar to the ordinary differential equations found for minimal surfaces by Hsiang and Lawson [9] in a more general setting. In particular, if $\dim M = 2$, the equations for harmonic maps are known to be closely related to the equations for minimal surfaces. We can actually study the minimal immersions

$$S^2 = S^1 \times R^1 \cup \{-\infty, \infty\} \longrightarrow S^k$$

and $S^1 \times S^1 = (S^1 \times R)/Z \longrightarrow S^k$ which are equivariant in the respect to an S^1 action. We are able to obtain families of immersed minimal 2-spheres in even dimensional spheres. These equivariant spheres are examples of the general families of spheres $S^2 \subset S^{2\ell}$ found by Calabi [5] and extensively studied by Chern [6] and Barbosa [3],[4].

The assumption of equivariance under a continuous group action whose orbits have co-dimension one in the domain manifold reduces a partial differential equation to an ordinary differential equation. The most interesting feature of this problem is the particular system of ordinary differential equations which arises. Moser has noticed that this system is the system of equations for the classical Neumann problem [11]. The author originally constructed a complete set of integrals for this problem in different form from the classical integrals [16]. This completely integrable system has been studied a great deal recently [2],[7],[10],[13],[15].

Gu has studied the Cauchy problem for Minkowski harmonic maps. He finds that solutions of the Sine Gordon equation may be obtained from harmonic maps $\Phi: E^{1,1} \to S^2$ [8]. In view of the known relationship between the Neumann problem and the Korteweg-deVries equation [2], [15], it might be fruitful to look further into the relationship between the Euclidean and Minkowski problems. See also T. Milnor [18].

Furthermore, the constructions of Calabi and Chern for the minimal 2-spheres are algebraic geometric in nature. Some investigation relating their construction of these special solutions to the Neumann problem and the algebraic properties of the minimal surface equation would be desirable.

Minimal volume maps $s: N \times R \to S^k$ which are equivariant with respect to a representation $\rho:G \to SO(k+1)$ can be found by studying a slightly different system of ordinary differential equations. The properties of this system are not understood for $n > 2$.

The main purpose of this article is to point out the existence of these questions. I greatly appreciate the continued interest and encouragement of J. Moser in this work.

Section §1. The Differential Equation

Let $M = N \times R$ where N is a compact symmetric space for the Lie group G. Assume $\rho:G \to SO(k+1)$ is given. In particular, we could take $N = S^{n-1}$ and $G = SO(n)$. Assume further that the set of equivariant maps.

$$F_\rho = \left\{s \in C^\infty(M,S^k): s \circ g = \rho(g)s \text{ for } g \in G\right\}$$

is non-empty. Select a basis $\{e_j\}_{j=1}^{\ell}$ of the Lie algebra of G and let $d\rho(1)e_j = A_j$. Then A_j is a $(k \times 1) \times (k \times 1)$ skew matrix. If we compute the energy of an equivariant maps $s:N \times [a,b] \to S^k$ we find

$$E(s) = \int_N d\mu \int_a^b \left(|s_t(t,\psi)|^2 + \sum_{j=1}^{\ell} |A_j(s(t,\psi))|^2\right)dt$$

for $\psi \in N$. Let $z(t) = s(t,\psi):[a,b] \to S^{k+1}$ for fixed $\psi \in N$.

Lemma 1.1: If $s \in F_\rho$ than s is harmonic if and only if $s(t,\psi) = z(t)$ is a critical map for the integral

$$(1) \qquad E(z) = \int_a^b \Big(|z'(t)|^2 + \sum_{i=1}^{\ell} |A_i z(t)|^2\Big)dt$$

for all $-\infty < a < b < +\infty$.

Proof: From general regularity theorems, the variational problem on $C^\infty(N \times [a,b],S^k)$ is equivalent to a variational problem on a Hilbert manifold $L^2_m(N \times [a,b],S^k)$. The general theory set forth by Palais [12] applies to the fixed point set of the map $s \to \rho(g)^{-1}s\circ g$. However, on the fixed point set $E(s) = \Big(\int_N d\mu\Big)\tilde{E}(z)$ and E and $\tilde{E}$ have corresponding critical points.

For notational convenience, note that $\sum_{i=1}^{\ell} A_i^2 = -\sum_{i=1} A_i A_i^* = -A^2$. Since the A_i are skew-symmetric, A may be taken as a non-negative symmetric matrix. Then equation (1) may be written as

$$(2) \qquad \tilde{E}(z) = \int_a^b \Big(|z'(t)|^2 + |Az(t)|^2\Big)dt \ .$$

Lemma 1.2: The Euler-Lagrange equations for an equivariant harmonic map into a sphere are

$$(3) \qquad z'' + \sum_{i=1}^{\ell} A_i^2 z + \lambda\, z = 0$$

or

$$(4) \qquad z'' - A^2 z + (|z'|^2 + |Az|^2)z = 0.$$

Proof: This is a standard Euler-Lagrange equation with the Lagrange multiplier $\lambda = \lambda(t)$ arising from the constraint $|z(t)| = 1$. The equation $\lambda = |z'|^2 + |Az|^2$ is computed by taking the dot product of (3) with z.

There are a number of standard integrals of this mechanical system. From conservation of energy we have

$$(5) \qquad |z'|^2 - |Az|^2 = I$$

Because of the original equivariance, the problem is invariant under the action of $e^{A_j\tau}$. From Noether's theorem we obtain angular momentum integrals

$$(6) \qquad (z' \cdot A_j z) = a_j.$$

Furthermore, if $e^{B\tau}\rho(G)e^{-B\tau} = \rho(G)$, or more generally if $[B,A^2] = 0$, we have additional angular momenta

$$(6') \qquad (z' \cdot Bz) = A_B.$$

The above integrals arise out of the symmetric structure of the problem. There is no reason a priori to expect more integrals. However, a complete system may be described as follows: Diagonalize the matrix A to have eigenvalues $\lambda_0 \leq \lambda_1 \leq \ldots . \leq \lambda_k$. Then, assuming $z = \{x_\alpha\}$ we have integrals (7) as well as angular momenta (6").

$$(6'') \qquad x_\alpha x'_\beta - x_\beta x'_\alpha = Q_{\alpha\beta}; \ \lambda_\alpha = \lambda_\beta .$$

$$(7) \qquad \sum_{\lambda_\beta = \lambda_\gamma} \left(\sum_{\lambda_\alpha \neq \lambda_\gamma} \frac{(x_\alpha x'_\beta - x_\beta x'_\alpha)^2}{\lambda_\beta^2 - \lambda_\alpha^2} - x_\beta^2 \right) = E_\gamma .$$

One can alway set $Q_{\alpha\beta} = 0$. This reduces the number of variables from $k+1$ to $\ell+1$ and confines the orbit to $S^\ell \subset S^k$. This yields the standard Neumann problem with $\lambda_\alpha \neq \lambda_\beta$ for $\alpha \neq \beta$. The integrals appear as

$$(8) \qquad \sum_{\beta\neq\alpha} \frac{(x_\alpha x'_\beta - x_\beta x'_\alpha)^2}{\lambda_\beta^2 - \lambda_\alpha^2} - x_\alpha^2 = E_\alpha \ .$$

See Al'ber [2], Devaney [7], Moser [10] or Ratiu [13] for more complete discussions of the additional "accidental" integrals (7-8).

§2. Equivariant minimal surfaces

If M has dimension 2, $M = S^1 \times R$, the connection between harmonic maps and minimal immersions is well-understood. In this case, $n = 1$ and A_1 is a $k\times k$ skew matrix with eigenvalues $\pm in_\alpha$ occuring in pairs except possibly for $n_0 = 0$. Rotate coordinat and eliminate any unnecessary ones:

$$(9) \qquad \begin{aligned} A_1 x_\alpha &= n_\alpha y_\alpha & 0 \leq n_\alpha \\ A_1 y_\alpha &= - n_\alpha x_\alpha ; & 0 \leq n_\alpha \end{aligned}$$

then $A^2 = - A_1^2$ has eigenvalues $\{n_\alpha^2\}$. The differential equations have the form

$$(10) \qquad \begin{aligned} x''_\alpha - n_\alpha^2 x_\alpha + \lambda x_\alpha &= 0 \\ y''_\alpha - n_\alpha^2 y_\alpha + \lambda y_\alpha &= 0 \ . \end{aligned}$$

$$\lambda = \sum_\alpha \left[(x'_\alpha)^2 + (y'_\alpha)^2 + n_\alpha^2 (x_\alpha^2 + y_\alpha^2) \right] = |z'|^2 + |Az|^2 \ .$$

We have automatically conservation of energy

$$(5) \qquad |z'|^2 - |Az|^2 = I$$

and conservation of angular momentum

(6) $$x'_\alpha y_\alpha - y'_\alpha x_\alpha = a_\alpha .$$

as well in the complete set including (7).

Lemma 2.1: Suppose $z(t) = \{x_\alpha(t), y_\alpha(t)\}$ and $s(e^{i\theta},t) = e^{A_1\theta} z(t)$. Then s is a conformal minimal immersion of $R^1 \times S^1$ if and only if z satisfies (10) and the integrals $I = 0$ and $\sum_\alpha n_\alpha a_\alpha = 0$.

Proof: Certainly s is harmonic, and it is a general fact that a harmonic map is a minimal immersion if and only if it is conformal [16]. However, s is conformal if

$$|s_t|^2 - |s_\theta|^2 = |z'|^2 - |Az|^2 = I = 0$$

(11)

$$(s_t \cdot s_\theta) = (z' \cdot A_1 z) = \sum_\alpha n_\alpha (x'_\alpha y_\alpha - y'_\alpha x_\alpha) = \sum_\alpha n_\alpha a_\alpha = 0 .$$

Of course it is not very interesting that there are a number of minimal immersions of $S^1 \times R \to S^k$. However, we can easily produce two types of immersions of compact surfaces from immersions of cylinders.

Proposition 2.2: Let z be a solution of (10) such that $s(t,\theta) = e^{A_1\theta} z(t)$ extends to a smooth minimal immersion of S^2. Then $I = a_\alpha = \alpha E_\alpha = 0$, and after eliminating unnecessary variables, $n_0 = 0$, $x_\alpha \not\equiv 0$, $0 \le \alpha \le \ell$, $x_\alpha \equiv 0$, $\alpha > \ell$ and $y_\alpha \equiv 0$.

Proof: In order for s to extend, $z(t)$ must approach fixed points of $e^{A_1\theta}$ as $t \to \pm\infty$. Then $n_0 = 0$, and the fixed points occur when $(x_0,y_0) = (\cos\theta, \sin\theta)$; $(x_\alpha,y_\alpha) = (0,0)$, $a > 0$. Write the second order equation (10) as a first order system

$$(12) \qquad z' = v \; ; \; v' = A^2 z - \lambda z .$$

Since $\lim_{t\to\pm\infty} z(t)$ exists, $\lim_{t\to\pm\infty} v(t) = 0$. However, if we evaluate the integrals on the fixed points of the system (11), $I = a_\alpha = E_\alpha = 0$. Because $a_\alpha = 0$, motion in the (x_α, y_α) plane occurs along a line and we may rotate coordinates so $y_\alpha = 0$. Order coordinates so $x_\alpha \neq 0$, $0 \leq \alpha \leq \ell$. The fact that $n_0 = 0$, $x_\alpha \neq 0$ follows from $\lim_{t\to\pm\infty} x_0(t) = \pm 1$.

Proposition 2.3: Let z be a solution of (13) on $-\infty < t < \infty$ such that $n_0 = 0$ and $|x_0(t)| \geq b > 0$ for $|t| \geq B$. Then s extends to a minimal immersion of $S^2 \to S^{2\ell} \subset S^k$.

Proof: It is sufficient to show $E(z) < \infty$, since it is true that $E(s) = 2\pi\tilde{E}(z) < \infty$ implies s extends to a minimal immersion [14]. Integrate the equation (10) in the first variable.

$$x_0''(t) + |z'(t)|^2 + A\, z(t)|^2\, x_0(t) = 0$$

For $T \geq B$,

$$|x_0'(T) - x_0'(B)| = \left| \int_B^T \left(|z'(t)|^2 + |A\, z(t)|^2\right) x_0(t)dt \right|$$

$$\geq b \int_T^T \left(|z'(t)|^2 + |A\, z(t)|^2\right)dt$$

However $|x_0'(T)| \leq |z'(T)| = \left(I + |A\, z(T)|^2\right)^{1/2}$ is bounded by $(I + n_\ell^2)^{1/2}$. So we have the inequality

$$\int_B^\infty (|z'(t)|^2 + |A\, z(t)|^2)dt \leq \frac{|x_0'(B)| + (I + n_\ell^2)^{1/2}}{b}$$

Exactly the same argument applies $-\infty < t \leq -B$.

This completes the proof.

We add here a remark on branch points. These immersions of $S^2 \to S^{2\ell}$ are unbranched except possibly at $\pm\infty$. Local analysis at the two critical points $(v = 0,\ z = (\pm 1,0))$ shows easily that an ℓ-dimensional family of solutions approach the points at $t \to +\infty$ and $t \to -\infty$. However, from this same local analysis, we see that, asymtotically $x_0 \simeq +1$, $x_\alpha \simeq c_\alpha e^{-n_\alpha|t|}$. Let the complex variable $w = e^{-|t|+i\theta}$. Then

$$s(t,\theta) = s(w) \simeq (1, c_1 w^{n_1}, c_2 w^{n_2}, \ldots, c_\ell w^{n_\ell}).$$

When $n_1 = 1$ the immersion is unbranched. Otherwise a branch point of order n_1 occurs.

We have shown that every S^1 equivariant $S^2 \subset S^k$ lies in $S^{2\ell} \subset S^k$. Here it corresponds to a solution of (10) with $y_\alpha \equiv 0$ and the integrals set

$$1 = E_0 = x_0^2 + \sum_{\beta>0} \frac{(x_0' x_\beta - x_\beta' x_0)^2}{n_\beta^2}$$

$$0 = E_\alpha = x_\alpha^2 + \sum_{\beta\neq\alpha} \frac{(x_\alpha' x_\beta - x_\beta' x_x)^2}{n_\beta^2 - n_\alpha^2} .$$

<u>Theorem 2.4</u>: Let $z(t)$ be a solution of (10) satisfying $n_0 = 0$ $a_\alpha = 0.\forall_\alpha$, $E_\alpha = 0$, $\alpha > 0$ and $E_0 = 1$.. Then $s(t,\theta) = e^{A_1\theta} z(t)$ extends to a smooth minimal immersion $S^2 \to S^{2\ell}$ with a branch point of order n_1 at the poles.

<u>Proof</u>: By Proposition 2.3 we need only show that $|x_0(t)| > b$ for $|t| > B$. In Lemma 2.6 we show that the angular momentum $x_\alpha' x_{\alpha-1} - x_\alpha x_{\alpha-1}'$ can not take on the value zero unless some coordinate is identically zero. Furthermore, using $E_\ell = 0$

$$x_\ell^2 = \sum_{\beta<\ell} \frac{(x'_\ell x_\beta - x'_\beta x_\ell)^2}{n_\ell^2 - n_\beta^2} .$$

Clearly $x_\ell(a) = 0$ implies each non-negative term in the sum is zero and the angular momentum $(x'_\ell x_{\ell-1} - x'_{\ell-1}x_\ell)(a) = 0$, which contradicts Lemma 2.6. So x_ℓ is never zero. Because the angular momentum $x'_\ell x_{\ell-1} - x'_{\ell-1}x_\ell$ is never sero, a zero of x_ℓ must occur between any two zeros of $x_{\ell-1}$. By induction we see that $x_{\ell-k}$ has at most k zeros. After the last zero of x_0, x_0 stays bounded away from zero.

Lemma 2.5: For $1 < \alpha \leq \ell$, the coordinate pair $(x_\alpha(a), x_{\alpha-1}(a)) = (0,$ implies one of the coordinates $(x_\alpha, x_{\alpha-1})$ is identically zero.

Proof: Manipulate the integrals (using $x_\alpha(a) = x_{\alpha-1}(a) = 0$)

$$x'^2_{\alpha-1}(a)\, E_\alpha - x'^2_\alpha\, E_{\alpha-1}$$

$$= \sum_{\substack{\beta<\alpha-1\\ \beta>\alpha}} (x'_{\alpha-1}x'_\alpha x_\beta)^2(a)\left(\frac{1}{n_\beta^2-n_\alpha^2} - \frac{1}{n_\beta^2-n_{\alpha-1}^2}\right)$$

$$= (x'_{\alpha-1}x'_\alpha)^2(a) \sum_{\substack{\beta<\alpha-1\\ \beta>\alpha}} x_\beta^2(a)\ \frac{(n_\alpha^2-n_{\alpha-1}^2)}{(n_\beta^2-n_\alpha)(n_\beta^2-n_{\alpha-1}^2)}$$

Assume $x'_\alpha(a)x'_{\alpha-1}(a) \neq 0$. If $\alpha > 1$, $E_\alpha = E_{\alpha-1} = 0$ and $x_\beta(a) = 0$ for all β, which violates the identity $\sum_\beta x_\beta^2(a) = 1$. If $\alpha = 0$, the left-hand side is negative, which is equally impossible. However, if $x'_\alpha(a)x'_{\alpha-1}(a) = 0$, one coordinate has both value and derivative zero at a, and so vanishes by the uniqueness theorem for O.D.E.

Lemma 2.6: If no coordinate x_β is identically zero, then $(x'_\alpha x_{\alpha-1} - x_\alpha x'_{\alpha-1})(a) \neq 0$ for all α, $\alpha,\beta \leq \ell$.

Proof: Assume $(x'_\alpha x_{\alpha-1} - x_\alpha x'_{\alpha-1})(a) = 0$. Then we may assume $x_\alpha(a) \neq 0$ and $x_{\alpha-1}(a) \neq 0$, since otherwise by uniqueness or by the previous lemma either x_α or $x_{\alpha-1}$ would be identically zero. Using the assumption, we manipulate the expression

$$x^2_{\alpha-1}(a)\, E_\alpha - x^2_\alpha(a)\, E_{\alpha-1}$$

until we find it equals

$$x^2_{\alpha-1}(a) \sum_{\substack{\beta<\alpha-1 \\ \alpha<\beta}} (x'_\alpha x_\beta - x'_\beta x_\alpha)^2(a)\; \frac{n^2_\alpha - n^2_{\alpha-1}}{(n^2_\beta - n^2_\alpha)(n^2_\beta - n^2_{\alpha-1})} \quad .$$

Every term in the sum is non-negative. The expression $x^2_{\alpha-1}(a)E_\alpha - x^2_\alpha(a)E_{\alpha-1}$ is zero if $\alpha > 1$ and $-x^2_1(a)$ for $\alpha = 1$. So when $(x'_{\alpha-1}x_\alpha - x'_\alpha x_{\alpha-1})(a) = 0$ we may conclude $(x'_\alpha x_\beta - 'x'_\beta x_\alpha)(a) = 0$ for all β. But $\sum_\beta (x'_\alpha x_\beta - x'_\beta x_\alpha)(a) x'_\beta(a) = -x_\alpha(a) \sum_\beta x'^2_\beta(a) = 0$. This is clearly impossible.

We now briefly describe the construction of harmonic maps of a torus into S^k. Every conformal torus is obtained from the flat cylinder $S^1 \times R$ by an identification $f_{\omega,\tau}(w,t) = (e^{i\omega}w, t+\tau) = (w,t)$ for $e^{i\omega} \in S^1$, $\tau \in R^t$. Let $T^2_{\omega,\tau} = (S^1 \times R)/[f_{\omega,\tau}]$.

Proposition 2.7: Let $z: R \to S^k$ be a solution of (3). If $z(t+\tau) = e^{A_1\omega} z(t)$, then the harmonic map $s: S^1 \times R \to S^k$ factors through the torus $T^2_{\omega,\tau}$:

$$S^1 \times R \longrightarrow T^2_{\omega,\tau} \xrightarrow{\ \hat{s}\ } S^k$$

Here $\hat{s}$ is harmonic. Furthermore $\hat{s}$ is a conformal (unbranched) minimal immersion if and only if the integrals (11) have value zero.

Proof: We must show $s(e^{i\omega} w, t+\tau) = s(w,t)$. But

$s(e^{i\omega}w,t+\tau) = s(e^{i\omega}e^{i\theta},t+\tau) = e^{A_1(\omega+\theta)}z(t,\tau) = e^{A_1\theta}z(t) = s(e^{i\theta},t)$ $= s(w,t)$ for $w = e^{i\theta}$. The second part follows from Lemma 2.1.

In view of the integrability of the system, one expects a number of solutions with this periodic property. It is somewhat more difficult to identify those which actually correspond to minimal tori. This problem has been treàted by Hsiang and Lawson when $S^k = S^3$ [9] in sufficient detail.

Explicit computations concerning area and curvature $K = -\dfrac{(\ln|Ax|^2)''}{|Ax|^2}$ can easily be obtained for both spheres and tori [16

§3. Minimal immersions of arbitrary dimension

For $n = \dim M > 2$, the harmonic maps $s: M \to S^k$ generally have little to do with minimal immersions. However, if we replace the energy integral by the modified integral

$$E_n(s) = \int_M |ds|^n \, d\mu$$

there is a relation. Assume the situation of Section 1. Then for equivariant maps $s \in F_\rho = \{s \in C^\infty(M,S^k) : s \cdot g = \rho(g)s \text{ for } g \in G\}$ where $M = N \times [a,b]$, we have again

$$E_n(s) = \int_N d\mu \int_a^b \left(|s_t(t,\psi)|^2 + \sum_j |A_j(s,\psi)|^2\right)^{n/2} dt$$

Setting $s(t,\psi) = z(t)$, we find that the corresponding integral for $z(t)$ is

$$E_n(z) = \int_a^b \left(|z'(t)|^2 + |Az(t)|^2\right)^{n/2} dt.$$

Now the Euler-Lagrange equations have the form

$$(\tilde{3}) \qquad (\rho z')' - \rho A^2 z + \rho \lambda z = 0$$

$$\rho(t) = (z'(t)|^2 + |Az(t)|^2)^{n/2-1}$$

$$(\tilde{4}) \qquad \lambda(t) = |z'(t)|^2 + |Az(t)|^2 \quad .$$

We still have conservation of energy

$$(\tilde{5}) \qquad \rho^{n/2}(|z'|^2 - |Az|^2) = I$$

and angular momentum

$$(\tilde{6}) \qquad (z' \cdot A_j z) = a_j .$$

Lemma 3.1: Suppose $z(t) = s(t,\psi)$, where $s \in F_\rho$. Then s is a minimal immersion if and only if z satisfies $(\tilde{3})$ and $I = 0$, $(z' \cdot A_j z) = a_j = 0$ all j.

Proof: Just as for $n = 2$, s is a minimal immersion if and only s is a conformal map which is a critical point of E_n [17]. Equivariant critical points of E_n correspond to critical points of $\tilde{E}_n$. The equations $I = 0$, $(z' \cdot A_j z) = 0$ are the equations of (weak) conformality.

We remark, without proving, that analogues of proposition 2.3 for $S^n = S^{n-1} \times R \cup \{\pm \infty\}$ and proposition 2.7 for $(N \times R)/f$ also apply. However, the construction of minimal tori and 2-spheres relies heavily on the existence of the integrals (6") and (7). Since no such system of integrals has been found for $(\tilde{3})$, the analysis of this case is not yet interesting. It is actually already quite computationally difficult to find the known isometric minimal immersions $S^n \subset S^k$ represented by solutions of this system of equations.

References

[1] R. Abraham and J. Marsden: Foundations of Mechanics, Second edition, Benjamin-Cummings (1978).

[2] S.I. Al'ber: On stationary problems for equations of Korteweg-de Vries type, Comm. Pure Appl. Math. 34 (1981), 259-272.

[3] J.L.M. Barbosa: On minimal immersions of S^2 into S^{2m}, Trans. Amer. Math. Soc. 210, (1975), 75-106.

[4] " : An extrinsic rigidity theorem for minimal immersions from S^2 into S^4, J. Diff. Geo. 14 (1979), 335-368.

[5] E. Calabi: Minimal immersions of surfaces in Euclidean spheres, J. Diff. Geo. 1 (1967), 111-125.

[6] S.S.Chern: On the minimal immersions of the two-sphere in a space constant curvature, Problems in Analysis, ed. Gunning, Princeton University Press (1970), 27-40.

[7] R. Devaney: Transversal homoclinic orbits in an integrable system, Amer. J. Math. 100 (1978), 631-642.

[8] Gu(Chao-Hao): On the Cauchy problem for harmonic maps defined on two-dimensional Minkowski space, Comm. Pure and Appl. Math 33, (1980), 727-738.

[9] W. Hsiang and H.B. Lawson: Minimal submanifolds of low cohomogene J. of Diff. Geometry 5 (1971), 1-38.

[10] J. Moser: Various aspects of integrable Hamiltonian systems, C.I. Bressone (1978), Progress in Mathematics 8, Birkhauser, Basel.

[11] C. Neumann: De problemate quodam mechanica, quad ad primam integralium ultra-ellipticowm classem revocatur, J. Reine und Angew. Math. 56 (1859), 54-66.

[12] R.S. Palais: The principle of symmetric criticality, Comm. Math. Phys. 69 (1979), 19-30.

[13] T. Ratiu: The C. Neumann problem as a completely integrable system on an adjoint orbit, Trans. Amer. Math. Soc. 264, (1981), 321-329.

[14] J. Sacks and K. Uhlenbeck: Minimal immersions of 2-spheres, Ann. of Math. 113 (1981), 1-24.

[15] E. Trubowitz: Lectures at the AMS summer conference (1979) and the New York Academy of Sciences (1979).

[16] K. Uhlenbeck: Minimal 2-spheres and tori in S^k, preprint (1975).

[17] " : Minimal spheres and other conformal variational problems, (to appear).

[18] T.K. Milnor: Characterizing harmonic immersions of surfaces with indefinite metric (preprint).

University of Illinois at Chicago Circle, Chicag
60680

Vol. 787: Potential Theory, Copenhagen 1979. Proceedings, 1979. Edited by C. Berg, G. Forst and B. Fuglede. VIII, 319 pages. 1980.

Vol. 788: Topology Symposium, Siegen 1979. Proceedings, 1979. Edited by U. Koschorke and W. D. Neumann. VIII, 495 pages. 1980.

Vol. 789: J. E. Humphreys, Arithmetic Groups. VII, 158 pages. 1980.

Vol. 790: W. Dicks, Groups, Trees and Projective Modules. IX, 127 pages. 1980.

Vol. 791: K. W. Bauer and S. Ruscheweyh, Differential Operators for Partial Differential Equations and Function Theoretic Applications. V, 258 pages. 1980.

Vol. 792: Geometry and Differential Geometry. Proceedings, 1979. Edited by R. Artzy and I. Vaisman. VI, 443 pages. 1980.

Vol. 793: J. Renault, A Groupoid Approach to C*-Algebras. III, 160 pages. 1980.

Vol. 794: Measure Theory, Oberwolfach 1979. Proceedings 1979. Edited by D. Kölzow. XV, 573 pages. 1980.

Vol. 795: Séminaire d'Algèbre Paul Dubreil et Marie-Paule Malliavin. Proceedings 1979. Edited by M. P. Malliavin. V, 433 pages. 1980.

Vol. 796: C. Constantinescu, Duality in Measure Theory. IV, 197 pages. 1980.

Vol. 797: S. Mäki, The Determination of Units in Real Cyclic Sextic Fields. III, 198 pages. 1980.

Vol. 798: Analytic Functions, Kozubnik 1979. Proceedings. Edited by J. Ławrynowicz. X, 476 pages. 1980.

Vol. 799: Functional Differential Equations and Bifurcation. Proceedings 1979. Edited by A. F. Izé. XXII, 409 pages. 1980.

Vol. 800: M.-F. Vignéras, Arithmétique des Algèbres de Quaternions. VII, 169 pages. 1980.

Vol. 801: K. Floret, Weakly Compact Sets. VII, 123 pages. 1980.

Vol. 802: J. Bair, R. Fourneau, Etude Géometrique des Espaces Vectoriels II. VII, 283 pages. 1980.

Vol. 803: F.-Y. Maeda, Dirichlet Integrals on Harmonic Spaces. X, 180 pages. 1980.

Vol. 804: M. Matsuda, First Order Algebraic Differential Equations. VII, 111 pages. 1980.

Vol. 805: O. Kowalski, Generalized Symmetric Spaces. XII, 187 pages. 1980.

Vol. 806: Burnside Groups. Proceedings, 1977. Edited by J. L. Mennicke. V, 274 pages. 1980.

Vol. 807: Fonctions de Plusieurs Variables Complexes IV. Proceedings, 1979. Edited by F. Norguet. IX, 198 pages. 1980.

Vol. 808: G. Maury et J. Raynaud, Ordres Maximaux au Sens de K. Asano. VIII, 192 pages. 1980.

Vol. 809: I. Gumowski and Ch. Mira, Recurences and Discrete Dynamic Systems. VI, 272 pages. 1980.

Vol. 810: Geometrical Approaches to Differential Equations. Proceedings 1979. Edited by R. Martini. VII, 339 pages. 1980.

Vol. 811: D. Normann, Recursion on the Countable Functionals. VIII, 191 pages. 1980.

Vol. 812: Y. Namikawa, Toroidal Compactification of Siegel Spaces. VIII, 162 pages. 1980.

Vol. 813: A. Campillo, Algebroid Curves in Positive Characteristic. V, 168 pages. 1980.

Vol. 814: Séminaire de Théorie du Potentiel, Paris, No. 5. Proceedings. Edited by F. Hirsch et G. Mokobodzki. IV, 239 pages. 1980.

Vol. 815: P. J. Slodowy, Simple Singularities and Simple Algebraic Groups. XI, 175 pages. 1980.

Vol. 816: L. Stoica, Local Operators and Markov Processes. VIII, 104 pages. 1980.

Vol. 817: L. Gerritzen, M. van der Put, Schottky Groups and Mumford Curves. VIII, 317 pages. 1980.

Vol. 818: S. Montgomery, Fixed Rings of Finite Automorphism Groups of Associative Rings. VII, 126 pages. 1980.

Vol. 819: Global Theory of Dynamical Systems. Proceedings, 1979. Edited by Z. Nitecki and C. Robinson. IX, 499 pages. 1980.

Vol. 820: W. Abikoff, The Real Analytic Theory of Teichmüller Space. VII, 144 pages. 1980.

Vol. 821: Statistique non Paramétrique Asymptotique. Proceedings, 1979. Edited by J.-P. Raoult. VII, 175 pages. 1980.

Vol. 822: Séminaire Pierre Lelong–Henri Skoda, (Analyse) Années 1978/79. Proceedings. Edited by P. Lelong et H. Skoda. VIII, 356 pages, 1980.

Vol. 823: J. Král, Integral Operators in Potential Theory. III, 171 pages. 1980.

Vol. 824: D. Frank Hsu, Cyclic Neofields and Combinatorial Designs. VI, 230 pages. 1980.

Vol. 825: Ring Theory, Antwerp 1980. Proceedings. Edited by F. van Oystaeyen. VII, 209 pages. 1980.

Vol. 826: Ph. G. Ciarlet et P. Rabier, Les Equations de von Kármán. VI, 181 pages. 1980.

Vol. 827: Ordinary and Partial Differential Equations. Proceedings, 1978. Edited by W. N. Everitt. XVI, 271 pages. 1980.

Vol. 828: Probability Theory on Vector Spaces II. Proceedings, 1979. Edited by A. Weron. XIII, 324 pages. 1980.

Vol. 829: Combinatorial Mathematics VII. Proceedings, 1979. Edited by R. W. Robinson et al.. X, 256 pages. 1980.

Vol. 830: J. A. Green, Polynomial Representations of GL_n. VI, 118 pages. 1980.

Vol. 831: Representation Theory I. Proceedings, 1979. Edited by V. Dlab and P. Gabriel. XIV, 373 pages. 1980.

Vol. 832: Representation Theory II. Proceedings, 1979. Edited by V. Dlab and P. Gabriel. XIV, 673 pages. 1980.

Vol. 833: Th. Jeulin, Semi-Martingales et Grossissement d'une Filtration. IX, 142 Seiten. 1980.

Vol. 834: Model Theory of Algebra and Arithmetic. Proceedings, 1979. Edited by L. Pacholski, J. Wierzejewski, and A. J. Wilkie. VI, 410 pages. 1980.

Vol. 835: H. Zieschang, E. Vogt and H.-D. Coldewey, Surfaces and Planar Discontinuous Groups. X, 334 pages. 1980.

Vol. 836: Differential Geometrical Methods in Mathematical Physics. Proceedings, 1979. Edited by P. L. García, A. Pérez-Rendón, and J. M. Souriau. XII, 538 pages. 1980.

Vol. 837: J. Meixner, F. W. Schäfke and G. Wolf, Mathieu Functions and Spheroidal Functions and their Mathematical Foundations Further Studies. VII, 126 pages. 1980.

Vol. 838: Global Differential Geometry and Global Analysis. Proceedings 1979. Edited by D. Ferus et al. XI, 299 pages. 1981.

Vol. 839: Cabal Seminar 77 – 79. Proceedings. Edited by A. S. Kechris, D. A. Martin and Y. N. Moschovakis. V, 274 pages. 1981.

Vol. 840: D. Henry, Geometric Theory of Semilinear Parabolic Equations. IV, 348 pages. 1981.

Vol. 841: A. Haraux, Nonlinear Evolution Equations- Global Behaviour of Solutions. XII, 313 pages. 1981.

Vol. 842: Séminaire Bourbaki vol. 1979/80. Exposés 543–560. IV, 317 pages. 1981.

Vol. 843: Functional Analysis, Holomorphy, and Approximation Theory. Proceedings. Edited by S. Machado. VI, 636 pages. 1981.

Vol. 844: Groupe de Brauer. Proceedings. Edited by M. Kervaire and M. Ojanguren. VII, 274 pages. 1981.

Vol. 845: A. Tannenbaum, Invariance and System Theory: Algebraic and Geometric Aspects. X, 161 pages. 1981.

Vol. 846: Ordinary and Partial Differential Equations, Proceedings. Edited by W. N. Everitt and B. D. Sleeman. XIV, 384 pages. 1981.

Vol. 847: U. Koschorke, Vector Fields and Other Vector Bundle Morphisms – A Singularity Approach. IV, 304 pages. 1981.

Vol. 848: Algebra, Carbondale 1980. Proceedings. Ed. by R. K. Amayo. VI, 298 pages. 1981.

Vol. 849: P. Major, Multiple Wiener-Itô Integrals. VII, 127 pages. 1981.

Vol. 850: Séminaire de Probabilités XV. 1979/80. Avec table générale des exposés de 1966/67 à 1978/79. Edited by J. Azéma and M. Yor. IV, 704 pages. 1981.

Vol. 851: Stochastic Integrals. Proceedings, 1980. Edited by D. Williams. IX, 540 pages. 1981.

Vol. 852: L. Schwartz, Geometry and Probability in Banach Spaces. X, 101 pages. 1981.

Vol. 853: N. Boboc, G. Bucur, A. Cornea, Order and Convexity in Potential Theory: H-Cones. IV, 286 pages. 1981.

Vol. 854: Algebraic K-Theory. Evanston 1980. Proceedings. Edited by E. M. Friedlander and M. R. Stein. V, 517 pages. 1981.

Vol. 855: Semigroups. Proceedings 1978. Edited by H. Jürgensen, M. Petrich and H. J. Weinert. V, 221 pages. 1981.

Vol. 856: R. Lascar, Propagation des Singularités des Solutions d'Equations Pseudo-Différentielles à Caractéristiques de Multiplicités Variables. VIII, 237 pages. 1981.

Vol. 857: M. Miyanishi. Non-complete Algebraic Surfaces. XVIII, 244 pages. 1981.

Vol. 858: E. A. Coddington, H. S. V. de Snoo: Regular Boundary Value Problems Associated with Pairs of Ordinary Differential Expressions. V, 225 pages. 1981.

Vol. 859: Logic Year 1979–80. Proceedings. Edited by M. Lerman, J. Schmerl and R. Soare. VIII, 326 pages. 1981.

Vol. 860: Probability in Banach Spaces III. Proceedings, 1980. Edited by A. Beck. VI, 329 pages. 1981.

Vol. 861: Analytical Methods in Probability Theory. Proceedings 1980. Edited by D. Dugué, E. Lukacs, V. K. Rohatgi. X, 183 pages. 1981.

Vol. 862: Algebraic Geometry. Proceedings 1980. Edited by A. Libgober and P. Wagreich. V, 281 pages. 1981.

Vol. 863: Processus Aléatoires à Deux Indices. Proceedings, 1980. Edited by H. Korezlioglu, G. Mazziotto and J. Szpirglas. V, 274 pages. 1981.

Vol. 864: Complex Analysis and Spectral Theory. Proceedings, 1979/80. Edited by V. P. Havin and N. K. Nikol'skii. VI, 480 pages. 1981.

Vol. 865: R. W. Bruggeman, Fourier Coefficients of Automorphic Forms. III, 201 pages. 1981.

Vol. 866: J.-M. Bismut, Mécanique Aléatoire. XVI, 563 pages. 1981.

Vol. 867: Séminaire d'Algèbre Paul Dubreil et Marie-Paule Malliavin. Proceedings, 1980. Edited by M.-P. Malliavin. V, 476 pages. 1981.

Vol. 868: Surfaces Algébriques. Proceedings 1976–78. Edited by J. Giraud, L. Illusie et M. Raynaud. V, 314 pages. 1981.

Vol. 869: A. V. Zelevinsky, Representations of Finite Classical Groups. IV, 184 pages. 1981.

Vol. 870: Shape Theory and Geometric Topology. Proceedings, 1981. Edited by S. Mardešić and J. Segal. V, 265 pages. 1981.

Vol. 871: Continuous Lattices. Proceedings, 1979. Edited by B. Banaschewski and R.-E. Hoffmann. X, 413 pages. 1981.

Vol. 872: Set Theory and Model Theory. Proceedings, 1979. Edited by R. B. Jensen and A. Prestel. V, 174 pages. 1981.

Vol. 873: Constructive Mathematics, Proceedings, 1980. Edited by F. Richman. VII, 347 pages. 1981.

Vol. 874: Abelian Group Theory. Proceedings, 1981. Edited by R. Göbel and E. Walker. XXI, 447 pages. 1981.

Vol. 875: H. Zieschang, Finite Groups of Mapping Classes of Surfaces. VIII, 340 pages. 1981.

Vol. 876: J. P. Bickel, N. El Karoui and M. Yor. Ecole d'Eté de Probabilités de Saint-Flour IX – 1979. Edited by P. L. Hennequin. XI, 280 pages. 1981.

Vol. 877: J. Erven, B.-J. Falkowski, Low Order Cohomology and Applications. VI, 126 pages. 1981.

Vol. 878: Numerical Solution of Nonlinear Equations. Proceedings, 1980. Edited by E. L. Allgower, K. Glashoff, and H.-O. Peitgen. XIV, 440 pages. 1981.

Vol. 879: V. V. Sazonov, Normal Approximation – Some Recent Advances. VII, 105 pages. 1981.

Vol. 880: Non Commutative Harmonic Analysis and Lie Groups. Proceedings, 1980. Edited by J. Carmona and M. Vergne. IV, 553 pages. 1981.

Vol. 881: R. Lutz, M. Goze, Nonstandard Analysis. XIV, 261 pages. 1981.

Vol. 882: Integral Representations and Applications. Proceedings. 1980. Edited by K. Roggenkamp. XII, 479 pages. 1981.

Vol. 883: Cylindric Set Algebras. By L. Henkin, J. D. Monk, A. Tarski, H. Andréka, and I. Németi. VII, 323 pages. 1981.

Vol. 884: Combinatorial Mathematics VIII. Proceedings, 1980. Edited by K. L. McAvaney. XIII, 359 pages. 1981.

Vol. 885: Combinatorics and Graph Theory. Edited by S. B. Rao. Proceedings, 1980. VII, 500 pages. 1981.

Vol. 886: Fixed Point Theory. Proceedings, 1980. Edited by E. Fadell and G. Fournier. XII, 511 pages. 1981.

Vol. 887: F. van Oystaeyen, A. Verschoren, Non-commutative Algebraic Geometry, VI, 404 pages. 1981.

Vol. 888: Padé Approximation and its Applications. Proceedings. 1980. Edited by M. G. de Bruin and H. van Rossum. VI, 383 pages. 1981.

Vol. 889: J. Bourgain, New Classes of $\mathcal{L}^p$-Spaces. V, 143 pages. 1981.

Vol. 890: Model Theory and Arithmetic. Proceedings, 1979/80. Edited by C. Berline, K. McAloon, and J.-P. Ressayre. VI, 306 pages. 1981.

Vol. 891: Logic Symposia, Hakone, 1979, 1980. Proceedings, 1979, 1980. Edited by G. H. Müller, G. Takeuti, and T. Tugué. XI, 394 pages. 1981.

Vol. 892: H. Cajar, Billingsley Dimension in Probability Spaces. III, 106 pages. 1981.

Vol. 893: Geometries and Groups. Proceedings. Edited by M. Aigner and D. Jungnickel. X, 250 pages. 1981.

Vol. 894: Geometry Symposium. Utrecht 1980, Proceedings. Edited by E. Looijenga, D. Siersma, and F. Takens. V, 153 pages. 1981.

Vol. 895: J.A. Hillman, Alexander Ideals of Links. V, 178 pages. 1981.

Vol. 896: B. Angéniol, Familles de Cycles Algébriques – Schéma de Chow. VI, 140 pages. 1981.

Vol. 897: W. Buchholz, S. Feferman, W. Pohlers, W. Sieg, Iterated Inductive Definitions and Subsystems of Analysis: Recent Proof-Theoretical Studies. V, 383 pages. 1981.

Vol. 898: Dynamical Systems and Turbulence, Warwick, 1980. Proceedings. Edited by D. Rand and L.-S. Young. VI, 390 pages. 1981.

Vol. 899: Analytic Number Theory. Proceedings, 1980. Edited by M.I. Knopp. X, 478 pages. 1981.